E-Textiles 2021: 3rd International Conference on the Challenges, Opportunities, Innovations and Applications in Electronic Textiles

E-Textiles 2021: 3rd International Conference on the Challenges, Opportunities, Innovations and Applications in Electronic Textiles

Editors

Steve Beeby

Kai Yang

Russel Torah

MDPI • Basel • Beijing • Wuhan • Barcelona • Belgrade

Editors
Steve Beeby
University of Southampton
UK

Kai Yang
University of Southampton
UK

Russel Torah
University of Southampton
UK

Editorial Office
MDPI
St. Alban-Anlage 66
4052 Basel, Switzerland

This is a reprint of Proceedings published online in the open access journal *Engineering Proceedings* (ISSN 2673-4591) in 2022 (available at: https://www.mdpi.com/2673-4591/15/1).

For citation purposes, cite each article independently as indicated on the article page online and as indicated below:

LastName, A.A.; LastName, B.B.; LastName, C.C. Article Title. *Journal Name* **Year**, *Article Number*, Page Range.

ISBN 978-3-0365-5475-4 (Pbk)
ISBN 978-3-0365-5476-1 (PDF)

Cover image courtesy of E-Textiles Network

Contents

About the Editors

Steve Beeby

Prof Steve Beeby holds a prestigious Royal Academy of Engineering Chair in Emerging Technologies on the topic of e-textile engineering. His research interests involve the application of flexible electronics, smart printable materials and energy-harvesting technologies for electronic textiles (e-textiles). He founded the E-Textiles Network and has over 300 publications and an h-Index of 50 with >15,000 citations. He is a co-founder of Perpetuum Ltd, a University spin-out formed in 2004 based upon vibration energy harvesting.

Kai Yang

Dr Kai Yang is an Associate Professor in Textile Design and Innovation. She is the Head of Research in Fashion and Textiles. Her research interests are electronic textiles, wearable healthcare and sustainable textiles. She has been awarded an EPSRC Fellowship and two MRC DPFS grants. She is the co-founder of Smart Fabric Inks, founder of Etexsense and co-chair of the E-textiles Network.

Russel Torah

Dr Russel Torah graduated with a BEng (hons) in Electronic Engineering and an MSc in Instrumentation and Transducers, both from the University of Southampton. In 2004, Russel obtained a PhD in Electronics from the University of Southampton. Since 2005, he has been a full-time researcher at the University of Southampton, where he is currently an Associate Professor. In 2011, Dr Torah co-founded Smart Fabric Inks Ltd, specialising in printed smart fabrics. His research interests are currently focused on smart fabric development, but he also has extensive knowledge of energy harvesting, sensors and transducers. Dr Torah has 145 publications, an h-index of 28 and 2 patents.

Preface to "E-Textiles 2021: 3rd International Conference on the Challenges, Opportunities, Innovations and Applications in Electronic Textiles"

The UK's E-Textile Network was established in 2018 with the goal of building the electronic textiles community and acting as a bridge between academia and industry. A key objective of the Network was to establish an international conference on the topic where the latest research and developments could be shared and disseminated. The E-Textiles conference series began in London in 2019, was held virtually in 2020, and for E-Textiles 2021, the conference was held in Manchester as a hybrid virtual and in-person event and included a mix of invited and accepted speakers from around the world. Accepted papers were peer-reviewed, and this collection of papers from the conference presents the latest work on a range of electronic textile applications and technologies. The Editors of the collection were responsible for organizing the conference with the assistance of the E-Textiles Network Steering Board and the Technical Programme Committee.

Steve Beeby, Kai Yang, Russel Torah
Editors

Editorial

Statement of Peer Review [†]

Russel Torah [1,*], Kai Yang [2] and Stephen Beeby [1]

1   School of Electronics and Computer Science, University of Southampton, Southampton SO17 1BJ, UK; spb@ecs.soton.ac.uk

2   Winchester School of Art, University of Southampton, Southampton SO17 1BJ, UK; ky2e09@soton.ac.uk

*   Correspondence: rnt@ecs.soton.ac.uk

†   Presented at the 3rd International Conference on the Challenges, Opportunities, Innovations and Applications in Electronic Textiles (E-Textiles 2021), Manchester, UK, 3–4 November 2021.

In submitting conference proceedings to *Engineering Proceedings*, the volume editors of the proceedings certify to the publisher that all papers published in this volume have been subjected to peer review administered by the volume editors. Reviews were conducted by expert referees to the professional and scientific standards expected of a proceedings journal.

- Type of peer review: double open reviews
- Conference submission management system: Web form + Excel
- Number of submissions sent for review: 32
- Number of submissions accepted: 30 (13 Oral, 17 Posters)
- Acceptance rate (number of submissions accepted/number of submissions received): 94% overall, 41% Oral.
- Average number of reviews per paper: 2
- Total number of reviewers involved: 12
- Any additional information on the review process: None.

Citation: Torah, R.; Yang, K.; Beeby, S. Statement of Peer Review. *Eng. Proc.* **2022**, *15*, 23. https://doi.org/10.3390/engproc2022015023

Published: 1 August 2022

engineering proceedings

Proceeding Paper

E-Textile Haptic Feedback Gloves for Virtual and Augmented Reality Applications †

Joash Chan and Russel Torah *

Electronics and Computer Science, University of Southampton, Southampton SO17 1BJ, UK; jwkc1g19@soton.ac.uk

* Correspondence: rnt@ecs.soton.ac.uk

† Presented at the 3rd International Conference on the Challenges, Opportunities, Innovations and Applications in Electronic Textiles (E-Textiles 2021), Manchester, UK, 3–4 November 2021.

Abstract: This paper outlines the development of e-textile haptic feedback gloves for virtual and augmented reality (VR/AR) applications. The prototype e-textile glove contains six Inertial Measurement Unit (IMU) flexible circuits embroidered on the fabric and seven screen-printed electrodes connected to a miniaturised flexible-circuit-based Transcutaneous Electrical Nerve Stimulator (TENS). The IMUs allow motion tracking feedback to the PC, while the electrodes and TENS provide electro-tactile feedback to the wearer in response to events in a linked virtual environment. The screen-printed electrode tracks result in haptic feedback gloves that are much thinner and more flexible than current commercial devices, providing additional dexterity and comfort to the user. In addition, all electronics are either printed or embroidered onto the fabric, allowing for greater compatibility with standard textile industry processes, making them simpler and cheaper to produce.

Keywords: e-textiles; haptic feedback glove; screen-printing; TENS; virtual reality; wearable

Citation: Chan, J.; Torah, R. E-Textile Haptic Feedback Gloves for Virtual and Augmented Reality Applications. *Eng. Proc.* **2022**, *15*, 1. https://doi.org/10.3390/engproc2022015001

Academic Editors: Steve Beeby and Kai Yang

Published: 8 March 2022

Publisher's Note: MDPI stays neutral with regard to jurisdictional claims in published maps and institutional affiliations.

1. Introduction

Haptic feedback gloves are wearable devices that provide a touch response to simulate tactile sensations of virtual objects [1]. Some methods employed to produce the haptics include force, vibrotactile, and thermal feedback—these are often very bulky or expensive to manufacture and limit the level of hand movements and natural feeling for the user. This research attempts an electro-tactile approach using TENS [2] due to the simplicity of the circuit required to achieve its signals, making it a suitable choice for an e-textile. The gloves aim to improve user interaction with VR/AR environments while also maintaining flexibility and breathability [3]. This study demonstrates the glove's functionalities through its interaction with a Graphical User Interface (GUI) via Bluetooth Low Energy (BLE).

2. Haptic Feedback Glove Control and System Design

The glove prototype consists of six IMUs (MPU6050—6 axes gyro + accelerometer) and six haptics channels, both located at the centre of the palm and on each fingertip. An additional electrode—the common ground of all channels—is located on the edge of the glove.

The IMUs communicate via the I2C protocol, but each IC has the same address, so the data line is controlled via a multiplexer (SN74LV4051A). A microcontroller sends data from each IMU to the GUI and controls the haptic stimulation when contact points in the GUI change.

TENS Circuitry

The TENS unit provides a tingling sensation to the user's skin at areas in contact with its electrodes by transmitting high-voltage electrical pulses. The circuit mainly consists of a boost converter (LT3467), a transistor (BC846BM3T5G) and a multi-channel high voltage analogue switch (MAX14866), as shown in Figure 1. This simplistic design minimises the overall circuit size.

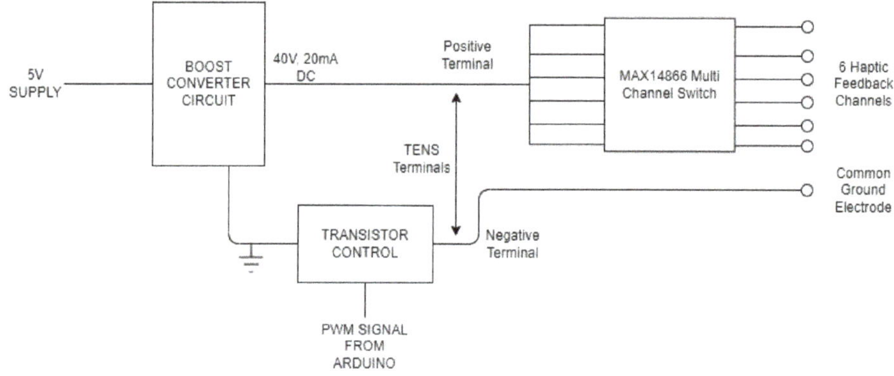

Figure 1. TENS unit system and control block diagram.

First, the boost converter generates a 40 V, 20 mA DC signal from a 5 V output of the microcontroller. A transistor is then connected between the ground and the negative terminal of the TENS output, while the 40 V output forms the stimulator's positive terminal. When the switching transistor is on, the negative terminal of the stimulator is grounded, giving rise to a 40 V potential between the two terminals. In contrast, zero potential exists between the terminals when the transistor is off. Therefore, the transistor is controlled by a PWM signal from the microcontroller to shape and generate the electrical pulses of the stimulator.

The negative terminal of the stimulator is connected to the common ground electrode, whereas the positive terminal is further connected to one end of each channel of the high-voltage, multi-channel analogue switch, allowing six independent haptic feedback channels.

The full 5 V to 40 V boost converter circuit diagram can be found in the LT3467 datasheet; the final strip circuit design layouts can be seen in Section 3.2.

3. E-Textile Fabrication

The IMU, TENS and control circuits are embedded into the glove via flexible 'strip circuits'—these are flexible circuits orientated in a strip form to aid integration into the textile [4]. Each strip design contains pads to solder copper Litz wires (embroidered on the glove), enabling connections with other circuits [5]. The IMU circuits are located on the fingers and the back of the hand; all other circuits are located on an extended wrist fabric attached to the glove.

3.1. Fabrication of Screen-Printed TENS Electrodes

Figure 2 shows the screen-printed electrode tracks on a polyester cotton fabric (Whaley's, Bradford, UK—OpticWhite Polyester/Cotton). These tracks are screen-printed to connect to the TENS to ensure the fabric in contact with the user's hand is continuous and flexible without impeding hand movement. The conductive silver ink tracks are printed between a 'primer', used to smooth the fabric where printed, and an 'encapsulation' coating, both of which use the same ink (Smart Fabric Inks Ltd., Southampton, UK—Fabinks UV-IF1004) [6]. The 'primer' consists of four printed layers providing a smooth base for the silver ink. The 'encapsulation' consists of two layers to electrically insulate the tracks from the user's skin and increase their durability.

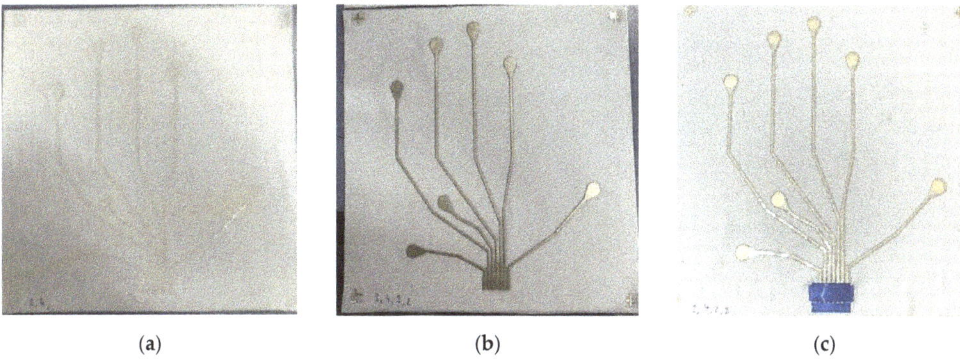

(a)   (b)   (c)

Figure 2. Screen-printed electrode tracks on polyester fabric: (**a**) 'Primer' layer; (**b**) Conductive layer printed on the 'primer'; (**c**) Final encapsulated electrode tracks with Amphenol Clincher crimped connector.

The conductive track is exposed at each end to allow contact. Hydrogel pads are attached to the rounded ends of each track to stimulate the fingertips and back of the hand. A crimped connector connects the printed tracks to the electrical stimulator.

3.2. Fabrication of Flexible Strip Circuits

The strip circuits are fabricated from copper laminated polyimide (Kapton) film (GTS Flexible Materials Ltd., Ebbw Vale, UK) due to its flexibility and thinness (25 µm), reducing its noticeability when integrated with the glove.

The conductive tracks are etched from the copper using a photolithography process developed previously at the University of Southampton [4] based on standard PCB processes. The strips were then cut out for component soldering.

Figure 3 shows examples of the circuit wafer and strip circuit. With single-layer copper, strip circuit designs are limited with no vias, creating significant design constraints. Hence, each strip design is simple, using embroidered wires for external connections. This e-textile fabrication technique allows for complex circuits to therefore be distributed around the garment, connecting to create more complex systems.

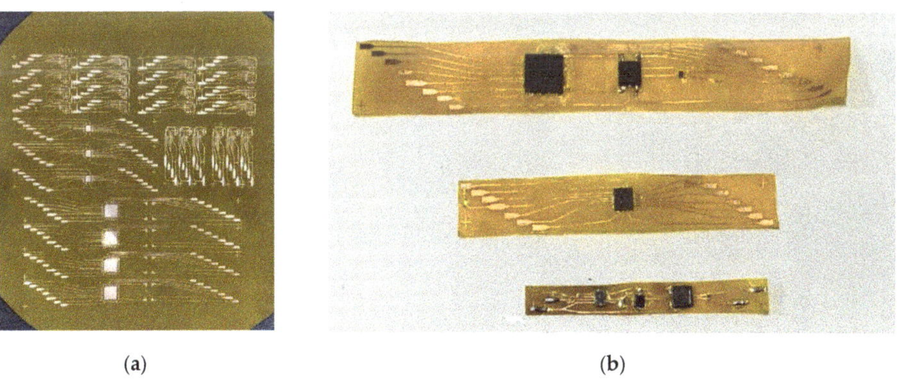

(a)   (b)

Figure 3. Electronic strip circuits: (**a**) Copper tracks on Kapton after etching stage; (**b**) Final strips with components soldered: high-voltage multiplexor (top), digital multiplexor (middle), boost converter (bottom).

4. Finished E-Textile Haptic Feedback Glove Prototype

The finished glove prototype is shown in Figure 4. The wrist extension attached to the glove contains an Arduino Nano 33 BLE, the TENS circuitries and a multiplexer connected to the IMUs. The IMU strip size is 24 × 5.7 mm; hence, they were easily embroidered onto the glove and can be fitted on each fingertip. All other strips are secured to the wrist with pockets stitched over them. A cotton glove is also stitched over the fabric of the IMU strips to be worn by the user. For this prototype, to make it easier to test, the electrodes were printed on the fabric separately to ensure that if one part failed, the entire e-textile was not lost. However, for future mass manufacturing, all of these parts could be combined.

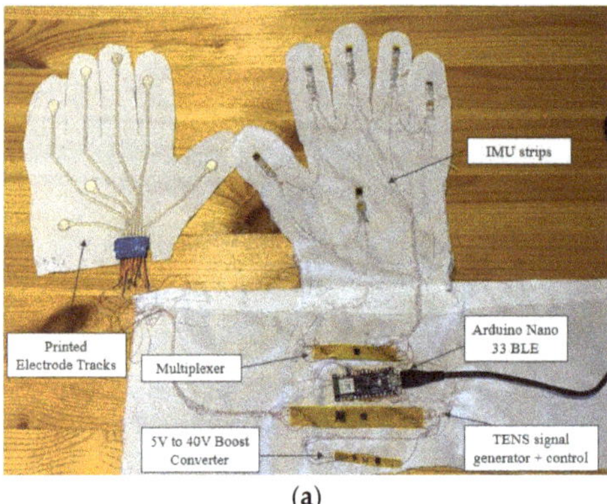

(a)

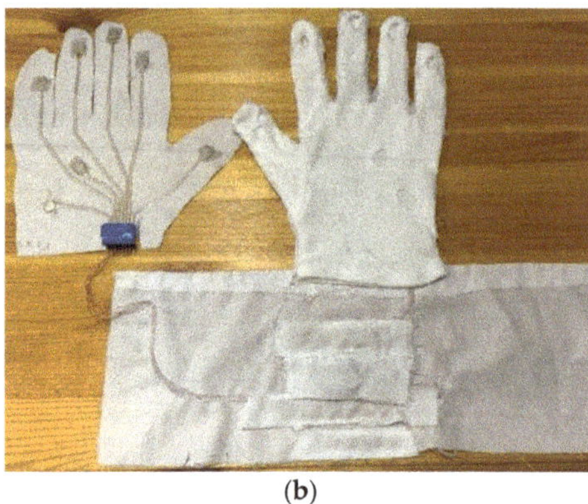
(b)

Figure 4. Finished haptic feedback glove prototype: (**a**) All electronics exposed for illustration; (**b**) Final glove design.

Figure 5 shows the functionalities of the glove. In this example, haptic feedback was provided to the wearer at the thumb and index finger channels as they are in contact with the sphere. The user's hand movements are tracked, and the haptic feedback is updated every frame to reflect events occurring in the GUI in real time.

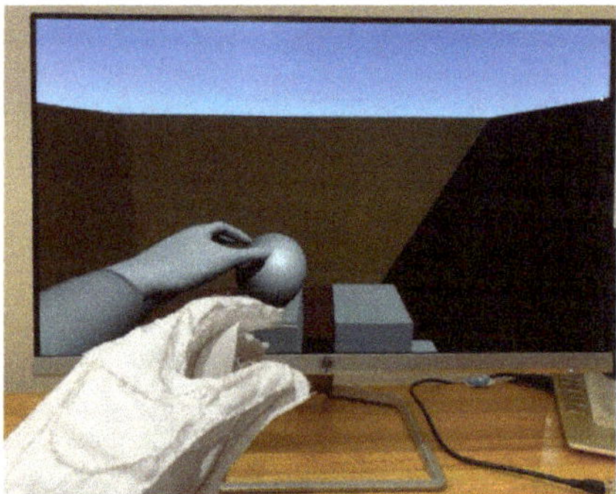

Figure 5. Demonstration of the e-textile glove's interaction with the GUI.

5. Conclusions

This paper demonstrates the development of simple and comfortable e-textile haptic feedback gloves using TENS. The technologies of screen printing and flexible strip circuits are integrated into the design, using embroidered wires to make any necessary connections. The strip circuits ensure small but robust e-textile circuitry while implementing screen-printed electrode tracks provides additional dexterity and comfort. Currently, the prototype can track wrist and finger movements and provide haptics at six channels, but these can be improved with additional IMUs and electrodes.

Future developments include reducing the size of ICs used to further reduce the overall circuit and replacing the electrodes with carbon electrodes printed directly on the glove along with their tracks. Additionally, multiple haptics patterns can be implemented by varying the stimulator pulse width and frequency to allow for different touch sensations.

Author Contributions: These authors contributed equally to this work. All authors have read and agreed to the published version of the manuscript.

Funding: This research was funded by the EU H2020 programme, grant number WEARPLEX—825339—wearplex.soton.ac.uk (accessed on 3 March 2022).

Institutional Review Board Statement: This study was conducted according to the guidelines of the Declaration of Helsinki and approved by the Ethics Committee of the University of Southampton (ERGO/FPSE/65891—approved 30 June 2021).

Informed Consent Statement: Informed consent was obtained from all subjects involved in the study and their permission to publish this paper.

Data Availability Statement: Not applicable.

Conflicts of Interest: The authors declare no conflict of interest.

References

1. Perret, J.; Benjamin, E. Touching Virtual Reality: A Review of Haptic Gloves. In Proceedings of the ACTUATOR 18, Bremen, Germany, 25–27 June 2018.
2. Kruijff, E.; Schmalstieg, D.; Beckhaus, S. Using neuromuscular electrical stimulation for pseudo-haptic feedback. In Proceedings of the ACM Symposium on Virtual Reality Software and Technology (VRST), Limassol, Cyprus, 1–3 November 2006.
3. Carlton, B. Smart Fabric Technology Brings Touch Haptics To The Oculus Quest. VR Scout, 27 October 2019. Available online: http://vrscout.com/news/smart-fabric-haptic-touch-oculus-quest/ (accessed on 25 June 2021).
4. Komolafe, A.; Torah, R.; Wei, Y.; Nunes-Matos, H.; Li, M.; Hardy, D.; Dias, T.; Tudor, M.; Beeby, S. Integrating flexible filament circuits for E-textile applications. *Adv. Mater. Technol.* **2019**, *4*, 1900176. [CrossRef]
5. Li, M.; Torah, R.; Nunes-Matos, H.; Wei, Y.; Beeby, S.; Tudor, J.; Yang, K. Integration and Testing of a Three-Axis Accelerometer in a Woven E-Textile Sleeve for Wearable Movement Monitoring. *Sensors* **2020**, *20*, 5033. [CrossRef] [PubMed]
6. Court, D.; Torah, R. Development of a Printed E-Textile for the Measurement of Muscle Activation via EMG for the Purpose of Gesture Control. In Proceedings of the International Conference on the Challenges, Opportunities, Innovations and Applications in E-Textiles (E-Textiles 2020), Online, 4 November 2020.

engineering proceedings

MDPI

Proceeding Paper

Finite-Element Analysis of the Mechanical Stresses on the Core Structure of Electronically Functional Yarns [†]

Mohamad Nour Nashed, Arash M. Shahidi, Theodore Hughes-Riley * and Tilak Dias

Advanced Textiles Research Group, Nottingham School of Art and Design, Nottingham Trent University, Dryden Street, Nottingham NG1 4GG, UK; m-nour.nashed@ntu.ac.uk (M.N.N.); arash.shahidi@ntu.ac.uk (A.M.S.); tilak.dias@ntu.ac.uk (T.D.)

* Correspondence: theodore.hughesriley@ntu.ac.uk; Tel.: +44-(0)1158-488178

† Presented at the 3rd International Conference on the Challenges, Opportunities, Innovations and Applications in Electronic Textiles (E-Textiles 2021), Manchester, UK, 3–4 November 2021.

Abstract: Electronic yarns (E-yarns) are a type of electronic textile where the electronics are embedded within the yarn structure, resulting in a yarn with normal textile properties. This is achieved by soldering thin copper wires onto electronic components, encapsulating the component within a UV-curable resin micro-pod, and covering the component with reinforcing yarns. The resin micro-pod protects both the component and the solder joints, providing the final yarn strength along its main axis: this is critical for its reliability after post-processing and during use. This work explored the use of Finite-Element Analysis to evaluate the mechanical stresses at the soldered joints of the core structure of E-yarns under axial loading before and after the encapsulation of the component. The results of this analysis were compared to the tensile test results of the core structure of the E-yarn.

Keywords: electronic textiles; smart textiles; electronic yarns; E-textiles; E-yarn; finite element analysis; electronically functional yarn; wearable electronics

check for **updates**

Citation: Nashed, M.N.; Shahidi, A.M.; Hughes-Riley, T.; Dias, T. Finite-Element Analysis of the Mechanical Stresses on the Core Structure of Electronically Functional Yarns. *Eng. Proc.* **2022**, *15*, 2. https://doi.org/10.3390/engproc2022015002

Academic Editors: Steve Beeby, Kai Yang and Russel Torah

Published: 9 March 2022

Publisher's Note: MDPI stays neutral with regard to jurisdictional claims in published maps and institutional affiliations.

1. Introduction

Developments in recent years have led to the miniaturization of the electronic devices that can be used in electronic textiles (E-Textiles) [1]. However, despite these advances, the reliability of the embedded electronics has proven to be a challenge for many E-textile innovations [1–3]. Textile garments are designed to be worn, washed, and dried, resulting in significant mechanical loads being applied to the fibres that form them [4]. The applied force can reach a few thousand Newtons per meter, which for E-textile garments can influence the reliability of any embedded electronic [5]. This lack of reliability is often due to contact failures or an increase in the resistance of the connections [2].

Electronic yarns (E-yarns) are a type of E-textile where small electronic components are embedded into the core of the yarns that make up the garment. The technology integrates the electronics by soldering the components onto copper wires. The functional part is then encapsulated within a rigid resin micro-pod to make it washable and increase the robustness of the solder joints [6]. This core structure (Figure 1a) is then inserted into a knitted or braided structure to provide the appearance of a normal textile yarn (Figure 1b). The process to manufacture E-yarns has been discussed in earlier publications [6,7].

Examining the failure mode of E-textiles is complicated, as stretching, bending, and indentation are all present in the life-cycle of a garment [8]. In the case of E-yarns, wash tests showed that the mode of failure was the breakage of the copper wire due to mechanical stresses [9]. The breakage of the copper wire was observed to occur at the interface between the rigid resin micro-pod and the conductive wire, as shown in Figure 1c. Significant damage to the copper wire may also lead to electrical failure prior to breakage. Thus, the E-yarn structure must be designed appropriately to avoid any premature failures.

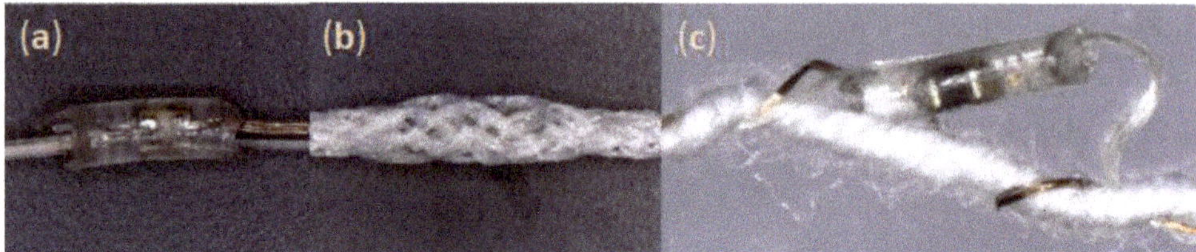

Figure 1. Microscope images of E-yarns. (**a**) The core structure of the E-yarn, including the conductive wire, the soldered component, and the supporting yarn. (**b**) The final E-yarn. (**c**) Image showing the failure of the conductive wire at the micro-pod interface after a wash test.

Different designs of E-yarn will use components with different geometries and the number and locations of the conductive wires in the core structure of the E-yarn depend on the component's connections. A different layout would lead to a different distribution of mechanical stresses and influence the failure mode. This study assessed the single lap joint soldering structure used to solder two-terminal surface mount devices such as LEDs and thermistors [6,7,9].

To understand the stresses on the core of the E-yarn, the structure was modelled using Finite-Element Analysis (FEA). A single-lap joint structure was used to assess the shear strength of the adhesives [10,11]. The mechanical loads transferred into the solder joints due to axial loading would result in shear stresses and bending moments that would result in the debonding of the solder pads of the component [12,13].

Understanding the distribution of mechanical stress within the core structure would help to improve the encapsulation process for the E-yarn and increase the reliability of the resulting yarn. This study will discuss the effect of adding the resin micro-pod around the electronic component on the mechanical stresses transferred into the solder joints of the core structure of the E-yarn.

2. Materials and Methods

2.1. Tensile Tests

The tensile results of the core structure were obtained using Zwick/Roell 2.5 tensile tester (Ulm, Germany) to evaluate the physical properties of the core structure before and after encapsulation (3 mm length micro-pod using Dymax 9001v3.5; Torrington, CT, USA). The tests were conducted using LEDs (Kingbright KPHHS-1005SURCK: New Taipei City, Taiwan) soldered onto copper wire (7-strand copper wire, 50 μm wire diameter: Knight Wire, Potters Bar, UK); 50 mm copper wire lengths and a test speed of 200 mm/min were used. The tests were conducted without a supporting yarn to evaluate the effect of adding only the micro-pod.

2.2. Computer Model of the Core E-Yarn Structure

A simplified 3D geometry was designed using Autodesk Inventor (Autodesk, Mill Valley, CA, USA). Models were prepared to simulate a $0.5 \times 0.5 \times 1.0$ mm component that was soldered onto a copper wire with a diameter of 0.15 mm. A second model represented the same soldered component encapsulated in a resin micro-pod. The 3D models were then used for finite-element static analysis using 'Ansys workbench 2021R1' (Ansys, Inc., Canonsburg, PA, USA). A static analysis was conducted to analyze the stress distribution in the single-lap joint of the electronic component soldered onto the copper wires. The analysis calculated the overall stresses at each element based on the von-mises criterion, shear stresses in the XZ (the plane parallel to the axis of the copper wire) direction, the highest stress in one of the principal directions at each element, and the highest shear stress at each element by plotting Mohr's circles at each element, using the principal stresses.

The following assumption was implemented to simplify the model:

- The solder joints' thickness was negligible.
- The copper wire was uniform and the physical properties of both the wire and the electronic component were homogenous.
- The applied load was axial to the wire's long axis, with deformation restricted in all other direction.
- Bonding between the conductive wire, the resin, and the contact was assumed to be perfect. No separation was assumed to occur under loading.
- The axial displacement of the fiber was assumed to be independent of the radius of the fiber.

3. Results

These tensile results showed that failure occurred 3.0 N before encapsulation, while after encapsulation failure occurred at 3.5 N, showing an improvement in the tensile strength after adding the resin micro-pod. Figure 2 shows tensile testing results for soldered LEDs before and after encapsulation.

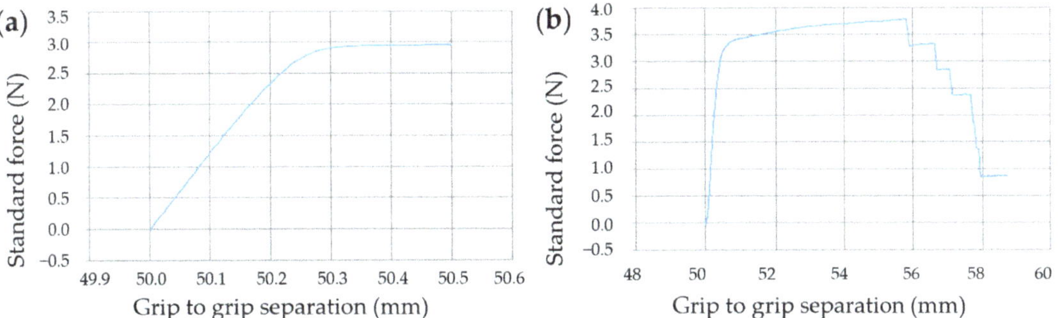

Figure 2. Tensile test results of the core structure of the E-yarn: (**a**) before encapsulation; (**b**) after encapsulation.

The results of Finite-Element Analysis are shown in Figure 3.

Results from the model showed that adding the resin micro-pod reduced the maximum stress to which the wire and soldered components were subjected. Moreover, it can be seen from the analysis that adding the resin micro-pod would reduce the equivalent stresses transferred to the solder joints by 5.8%, in comparison to the stresses before encapsulation, while the maximum shear stresses were similarly reduced by 6.8%. These results are in line with empirical testing, which showed an increase in the total strength after adding the micro-pod to the core structure of the E-yarn.

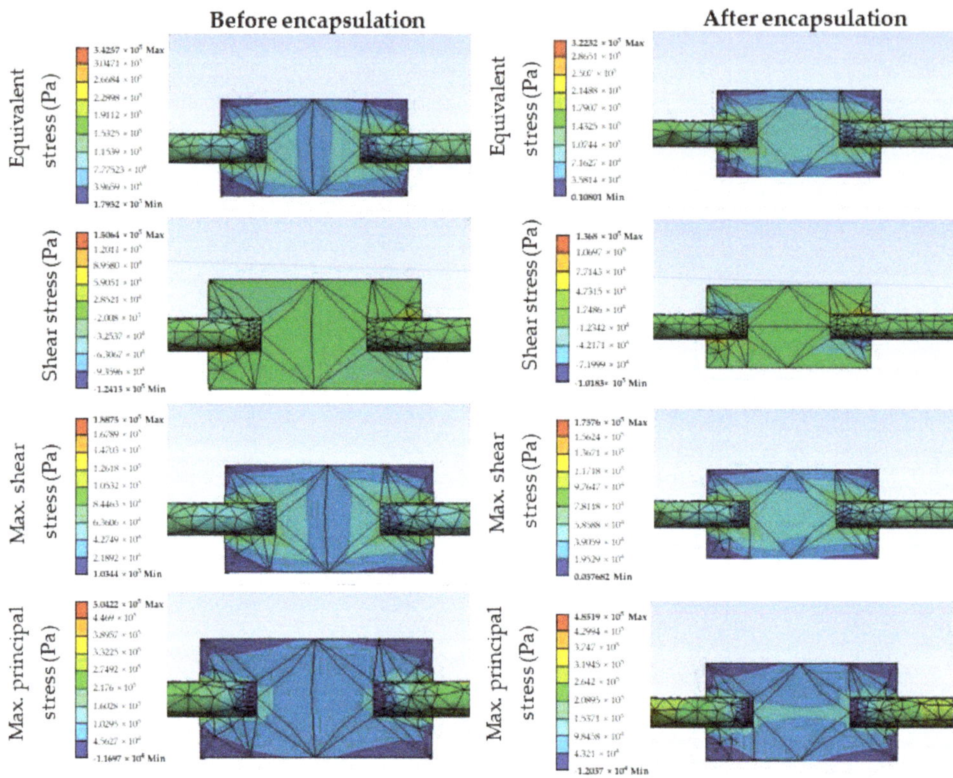

Figure 3. The results of the FEA for the core structure of the E-yarn. The figure shows the bottom view of the soldered components. Note that the micro-pod has been hidden for clarity.

4. Discussion and Conclusions

Adding a resin micro-pod to the single-lap joint structure resulting from the process of soldering a copper wire onto a component changed the distribution of stresses in the matrix. Failure of the interfacial bonds occurred when the shear strength of the structure, or the tenacity of the wire, are reached. The analysis showed that the maximum shear stress was near the ends of the solder joints. The tensile results showed an increase in the strength of the core structure by 16% in comparison to that before encapsulation. The FEA analysis results supported this.

FEA analysis is suitable for predicting the mechanical failure in E-yarns and can be extended for different arrangements and sizes of the component. It can also be used to evaluate different types of resin.

Author Contributions: Conceptualization, M.N.N. and T.D.; methodology, M.N.N. and A.M.S.; formal analysis, M.N.N.; investigation, M.N.N. and A.M.S.; writing—original draft preparation, M.N.N. and T.H.-R.; writing—review and editing, A.M.S. and T.D.; supervision, T.H.-R. and T.D.; funding acquisition, T.H.-R. and T.D. All authors have read and agreed to the published version of the manuscript.

Funding: This research was funded by the Engineering and Physical Sciences Research Council (EPSRC) grant EP/T001313/1, entitled 'Production engineering research for the manufacture of novel electronically functional yarns (E-yarn) for multifunctional smart textiles' and EPSRC grant number EP/M015149/1, entitled 'Novel manufacturing methods for functional electronic textiles'.

Institutional Review Board Statement: Not applicable.

Informed Consent Statement: Not applicable.

Data Availability Statement: The datasets used in this study are available on reasonable request.

Acknowledgments: The researcher would like to thank CARA for their support during the project.

Conflicts of Interest: The authors declare no conflict of interest.

References

1. Hughes-Riley, T.; Dias, T.; Cork, C. A Historical Review of the Development of Electronic Textiles. *Fibers* **2018**, *6*, 34. [CrossRef]
2. Tao, X.; Koncar, V.; Huang, T.-H.; Shen, C.-L.; Ko, Y.-C.; Jou, G.-T.; Tao, X.; Koncar, V.; Huang, T.-H.; Shen, C.-L.; et al. How to Make Reliable, Washable, and Wearable Textronic Devices. *Sensors* **2017**, *17*, 673. [CrossRef] [PubMed]
3. Virkki, J.; Björninen, T.; Kellomäki, T.; Merilampi, S.; Shafiq, I.; Ukkonen, L.; Sydänheimo, L.; Chan, Y.C. Reliability of washable wearable screen printed UHF RFID tags. *Microelectron. Reliab.* **2014**, *54*, 840–846. [CrossRef]
4. De Vries, H.; Peerlings, R. Predicting conducting yarn failure in woven electronic textiles. *Microelectron. Reliab.* **2014**, *54*, 2956–2960. [CrossRef]
5. Crow, R.M.; Dewar, M.M. Stresses in Clothing as Related to Seam Strength. *Text. Res. J.* **1986**, *56*, 467–473. [CrossRef]
6. Hardy, D.A.; Anastasopoulos, I.; Nashed, M.-N.; Oliveira, C.; Hughes-Riley, T.; Komolafe, A.; Tudor, J.; Torah, R.; Beeby, S.; Dias, T. Automated insertion of package dies onto wire and into a textile yarn sheath. *Microsyst. Technol.* **2019**, 1–13. [CrossRef]
7. Nashed, M.-N.; Hardy, D.A.; Hughes-Riley, T.; Dias, T. A Novel Method for Embedding Semiconductor Dies within Textile Yarn to Create Electronic Textiles. *Fibers* **2019**, *7*, 12. [CrossRef]
8. De Kok, M.; De Vries, H.; Pacheco, K.; Van Heck, G. Failure modes of conducting yarns in electronic-textile applications. *Text. Res. J.* **2015**, *85*, 1749–1760. [CrossRef]
9. Hardy, D.A.; Rahemtulla, Z.; Satharasinghe, A.; Shahidi, A.; Oliveira, C.; Anastasopoulos, I.; Nashed, M.N.; Kgatuke, M.; Komolafe, A.; Torah, R.; et al. Wash Testing of Electronic Yarn. *Materials* **2020**, *13*, 1228. [CrossRef] [PubMed]
10. Da Silva, L.F.M.; Lima, R.F.T.; Teixeira, R.M.S.; Puga, A. Closed-Form Solutions for Adhesively Bonded Joints: Two-Dimensional Linear Elastic Analyses. 2008. Available online: https://paginas.fe.up.pt/~{}em03108/Report1_Closed_form_models.pdf (accessed on 10 December 2021).
11. Da Silva, L.F.M.; das Neves, P.J.C.; Adams, R.D.; Spelt, J.K. Analytical models of adhesively bonded joints-Part I: Literature survey. *Int. J. Adhes. Adhes.* **2009**, *29*, 319–330. [CrossRef]
12. Han, X.; Crocombe, A.D.; Anwar, S.N.R.; Hu, P. The strength prediction of adhesive single lap joints exposed to long term loading in a hostile environment. *Int. J. Adhes. Adhes.* **2014**, *55*, 1–11. [CrossRef]
13. Gustafsson, J. Stress Equation for 2D Lap Joints with a Compliant Elastic Bond Layer. 2008. Available online: https://www.byggmek.lth.se/fileadmin/byggnadsmekanik/publications/tvsm7000/web7148.pdf (accessed on 10 December 2021).

engineering proceedings

Proceeding Paper

Development of Smart Kneecap with Electrical Stimulation †

Jitheesh V R [1],*, Rashmi Thakur [2] and Prabir Jana [1]

[1] Department of Fashion Technology, National Institute of Fashion Technology, New Delhi 110016, India; prabir.jana@nift.ac.in
[2] Department of Fashion Technology, National Institute of Fashion Technology, Mumbai 410210, India; rashmi.thakur@nift.ac.in
* Correspondence: jitheesh.vr@gmail.com
† Presented at the 3rd International Conference on the Challenges, Opportunities, Innovations and Applications in Electronic Textiles (E-Textiles 2021), Manchester, UK, 3–4 November 2021.

Abstract: This research was conducted to develop a textile electrode-based TENS module, as an alternative to overcome the shortcomings of current conventional electrodes used in TENS enabled pain therapy products, such as kneecaps. The existing TENS devices were studied to carry out a comparative analysis of their features and shortcomings. Three sets of textile electrodes were developed using conductive yarn knitted structure, conductive yarn embroidered fabric, and coated conductive fabric. In addition, the smart kneecap was developed with the aim of providing an electrical stimulation to the knee using a TENS module. The TENS module was actuated using a switching circuit of MOSFET and textile electrodes for the smooth conduction of electric stimulation into the body. These electro stimulations have been proven to be helpful in relieving the pain in the knee, as well as in giving the knee, thigh, and leg a sense of relaxation. Testing was conducted and subjective feedback was collected for the developed prototype.

Keywords: transcutaneous electrical nerve stimulation (TENS); textile electrode; smart wearables; conductive textile; kneecap

Citation: Jitheesh V R; Thakur, R.; Jana, P. Development of Smart Kneecap with Electrical Stimulation. *Eng. Proc.* **2022**, *15*, 3. https://doi.org/10.3390/engproc2022015003

Academic Editors: Steve Beeby, Kai Yang and Russel Torah

Published: 9 March 2022

Publisher's Note: MDPI stays neutral with regard to jurisdictional claims in published maps and institutional affiliations.

1. Introduction

Transcutaneous electrical nerve stimulation (TENS) is a method of electrical stimulation, which primarily aims to provide pain relief. It works on the principle of exciting the sensory nerves, and stimulating the pain gate mechanism. TENS device is a safer mode of treatment without any side effects [1]. TENS knee brace provides pain relief and muscle relaxation by stimulating the muscles and motor points, enabling better blood circulations. This research work utilized the concept of TENS therapy for pain and muscle relaxation. Using this concept, a handheld TENS module device was developed using textile electrodes, unlike conventional carbon rubber and hydrogel electrodes.

An electric circuit integrated with MOSFET was developed for the generation of electric pulses, which was controlled via a microcontroller. This helped in the generation of symmetrical biphasic waveforms. The frequency and pulse width of waveforms were adjusted by changing the variables in the microcontroller program. The intensity of electric pulses was adjusted by varying the input voltage.

TENS modules used in kneecaps have been found to be efficient, noninvasive, and with no side effects for overcoming chronic pains. However, the existing modules are found to be uncomfortable. This was explored further to realize the shortcomings of the current existing electrodes used in TENS modules, such as hydrogels and carbon rubber electrodes. Three types of frequency modes are used in TENS. (1) High frequency (90–130 Hz), which initiates the pain gate mechanism by activating the A beta sensory nerves thus reducing the transmission of pain stimulus to reach the brain (2) Low frequency (2–5 Hz), which activates the opioid mechanism and causes the release of an endogenous opiate in spinal cord. This acts as a pain killer and helps in relieving the pain [2]. (3) Burst frequency is the third variant, which is a combination of high and low frequencies [3]. Any electric stimulation using the skin surface electrode with the intention of stimulating nerves that provide symptomatic pain relief can be referred to as TENS [4]. The burst frequency activates both mechanisms and provides an effective result. Electrodes are used to conduct these electric pulses into the body. The positioning of the electrodes is important. Electric stimulation works effectively if placed appropriately at the required location of the body. It should be near the location where the highest intensity of pain exists and near the nerve trunk trigger position in order to block the signal's transmission to the brain. Notably, the electrode needs to be attached to the proper dermatome.

Textile electrodes were developed to overcome the existing drawbacks of conventional electrodes using conductive fabric, as well as integrating conductive yarns through knitting and embroidery. Four different conductive fabrics were selected for the development of textile electrodes, which resulted in four different kneecaps as the final prototype. The final testing and user feedback were considered to evaluate the performance of the TENS module and textile electrode performance.

2. Materials and Methods

2.1. TENS Frequency

Herein, it is possible to give a low frequency range or high frequency range, depending on the end user's requirements, although the present study has utilized the burst frequency methodology for the TENS module. In addition, the module can be controlled via a microcontroller.

2.2. Comparative Analysis

A comparative analysis was conducted by studying the commercially existing TENS products while considering the electrode. The summary of same is provided in Table 1. The textile electrodes were found to overcome major drawbacks of the conventional electrodes. One shortcoming of textile electrode is its reduction in conductivity when dry, which states that wet electrodes will be more comfortable than dry electrodes [5,6]. In addition, they are found to be durable, washable, breathable, and flexible [7]. Therefore, the textile electrode was selected for the study.

Table 1. Comparison of different electrodes.

Electrode Category	Functional Advantages	Shortcomings
Hydrogel electrodes	• Adhesive property • No external conductive gel solution is required	• Hygiene problem owing to the gel used • Adhesive property is lost after repeated usage, thus requiring frequent replacement • Skin irritation
Carbon rubber electrodes	• Durable • Repetitive use is possible	• Conductive gel may cause some irritation • Less conductive • Higher thickness
Textile electrodes	• Durable • More flexible • Thinner and attached/adhered to apparels, such as wearables • More conductive	• Textile electrode has to be moisturized for dry skin types, as dry skin offers less conductivity

2.3. Electrodes

Three methods were used to manufacture the textile electrodes. First, the electrode was developed by knitting using conductive yarns integrated with nonconductive yarns. Second, the conductive yarn was embroidered to make the fabric conductive and utilize it as an electrode. Finally, the conductive coated fabric was used to develop the textile electrode. Based on the comparative analysis one of the method was selected for further studies.

2.3.1. Conductive Yarn Knitted Electrode

Conductive yarns are used to create a 1×1 square size patch knitted with nonconductive yarn. A single jersey (10 gauge fabric) is knitted in a hand flat machine. The conductive yarn used was a blend of 20% stainless steel and 80% polyester, of English Count (Ne) 28/2. The wales per inch (wpi) is 19 and the courses per inch (cpi) is 21 (Figure 1).

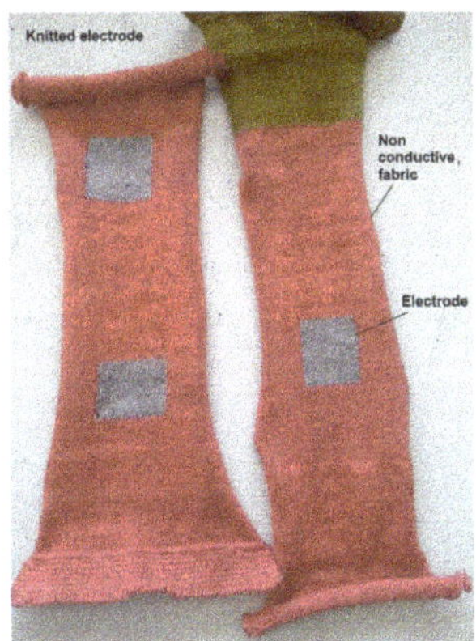

Figure 1. Knitted electrode.

2.3.2. Conductive Yarn Embroidered Electrode

Here, the conductive yarn (20% stainless steel and 80% polyester) of English count (N_e) 28/2 was used for embroidery. Two 1×1 square inch shape patches were embroidered on a 100% polyester base fabric. The embroidery machine speed was 1000 stitches per minute (SPM) and the number of stitches per square inch was 1025 (indicates less density) and 1710 (indicates more density), respectively. Both patches were tested separately, but gave similar results. Images of both the patches are given in Figure 2.

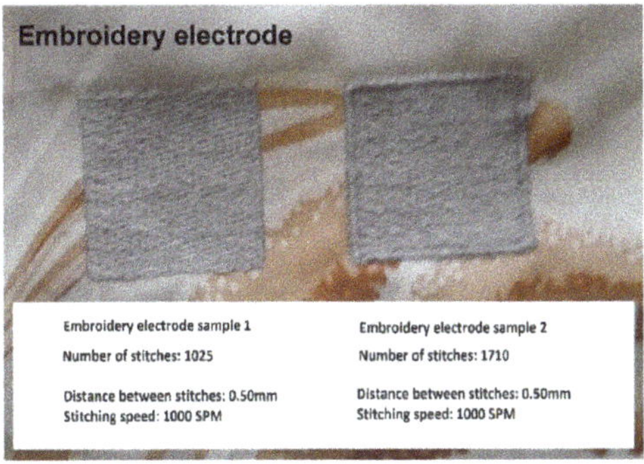

Figure 2. Embroidered electrode patches.

2.3.3. Fabric with Conductive Coating

This textile electrode was constructed by a conductive fabric patch with a silver coating of 99.9%. The patch was attached to the Velcro strap, as shown in Figure 3.

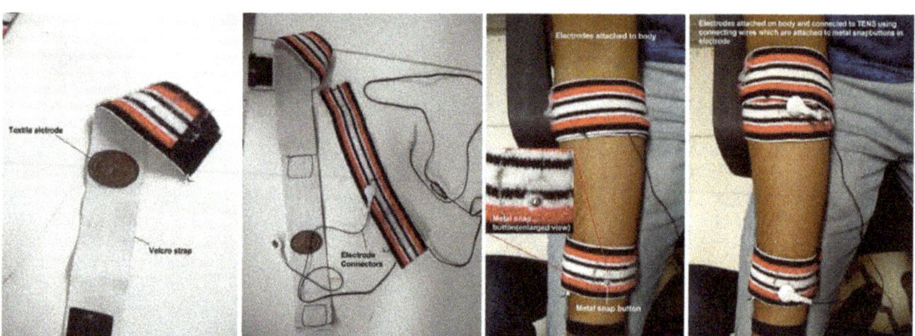

Figure 3. Sample textile electrode.

By improving the conductive properties of textile material, the electrode impedance can be reduced, thus improving polarizability. This further increases in conductivity, which leads to an improvement in the efficiency of TENS module [8].

2.4. Prototype Development

The prototype consists of a TENS module with textile electrodes attached to the kneecap. TENS module is connected to the electrodes in the kneecap using wires and a metal snap button. These kneecaps can be worn on the knee. In this way, electric pulses generated by the TENS module will be transferred to the skin via electrodes. A schematic representation of the module is shown in Figure 4.

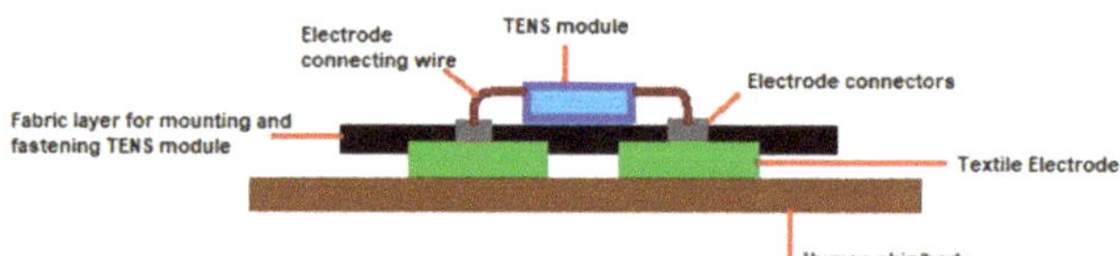

Figure 4. Smart kneecap with TENS structure.

2.4.1. Final Textile Electrodes Development

The knitted textile electrode, prepared through knitting of the conductive yarn, was found to be uncomfortable during testing due to irritation and a tingling effect. Furthermore, both sides of the electrode were exposed. As a result, this led to the chance of experiencing an electrical pulse on the external side. Moreover, it involved the high cost of manufacturing.

The textile electrode, prepared through conductive yarn embroidery, was found to be aesthetically appealing with a softer feel and lower manufacturing cost. However, irritation and a tingling effect were reported, as well. Furthermore, with repetitive and prolonged usage, the chances of fraying around the edges of the embroidered conductive yarn were anticipated.

Herein, it was observed that conductive coated fabric patches were the most efficient and suitable textile electrode among all of the textile electrodes developed. Therefore, this category of textile electrode was selected for further prototype development. Four different conductive coated fabrics were sourced for developing the textile electrodes. The commercial names have been replaced and coded as brands P, Q, R, S for confidentiality.

The specification and properties of these fabrics are furnished in Table 2. The sample textile electrode given in Figure 3 was prepared using the 'P' fabric.

Table 2. Specification and properties of conductive fabrics used as an electrode (data published by brands).

Name of Fabric	P	Q	R	S
Description	Stretch conductive knitted fabric coated with medical-grade silver	Silver (Ag) plated knitted fabric	Silver plated knitted fabric	polyester copper and nickel woven
Coating	99.9% silver coating	Plating: 99% pure silver coating (+B): Polyurethane as additional protective coating	plated with real silver	Nickel-copper-nickel-plated
Fiber Composition	Not disclosed	94% nylon + 6% elastomer	83% nylon + 17% silver	Polyester-based woven cloth
Fabric Thickness	0.40 mm	0.022″ (0.56 mm) ± 10%	0.50 mm	0.1 mm
Operational Temperature Range	−30 to 90 °C	−30 to 90 °C/−22 to 194 °F	−30 to 90 °C	−30 to 90 °C
Surface resistivity	<0.5 Ohm/sq. (unstretched)	<2 Ω/sq. (front/visible side)	1 ohm per foot in any direction	0.05 to 0.1 Ω/sq

The fabrics were cut into rectangular pieces (size 7 × 4 cm) and attached as a double layer to the kneecap in suitable positions. The metal snap button was used to connect the electrode with the external wires. Positions were selected near the motor points located on the upper side of the knee (Figure 5). These electrodes were attached to the respective positions in the kneecap (Figure 6). A Velcro strap was attached to the external side of the kneecap, which can be wrapped around the kneecap for better fastening of electrodes. Similarly, all of the four fabric patches were attached to different kneecaps resulting in four variants, as shown in Figure 7.

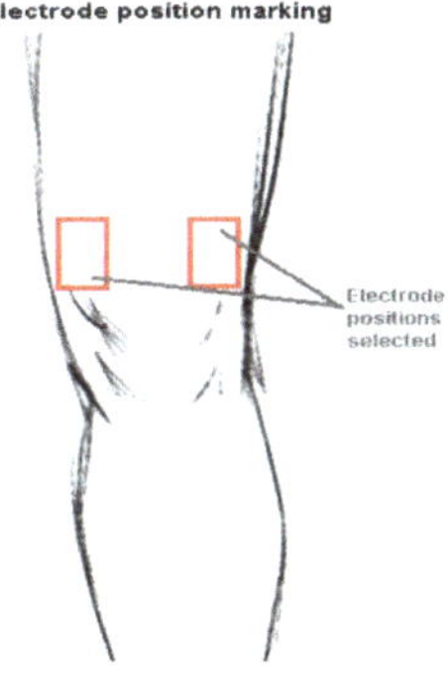

Figure 5. Electrode position selection for attachment in the kneecap.

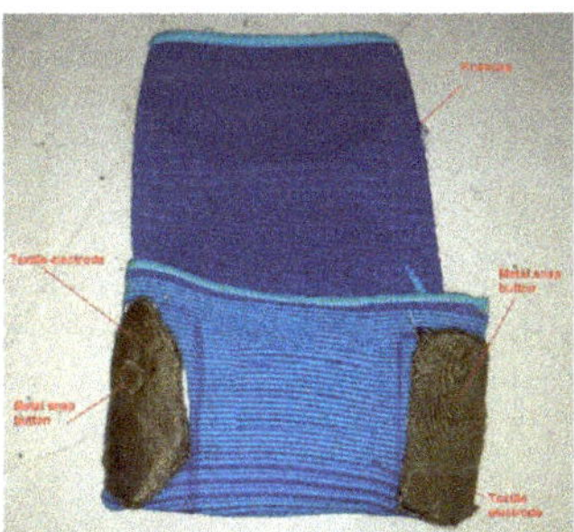

Figure 6. Textile electrode attached to the inside of kneecap.

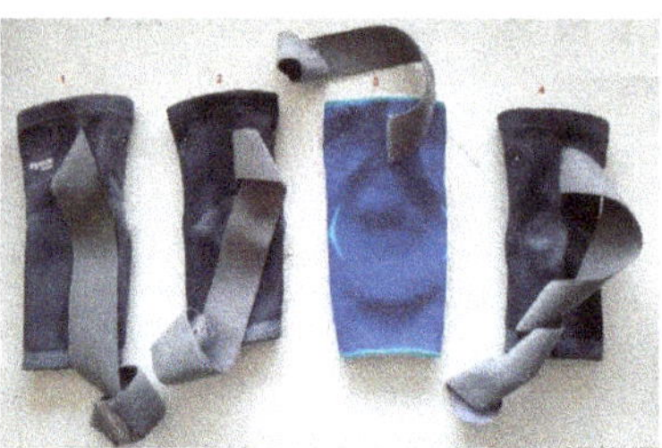

Figure 7. Developed kneecaps with attached textile electrodes.

2.4.2. TENS Module Development

A MOSFET switching circuit that works on the H-bridge concept was used for electric pulse generation, which was controlled through an Arduino Uno microcontroller. The frequency and pulse width were adjusted through microcontroller programming. The power supply was drawn from a rechargeable 3.3 V LG MJ1 18650 3500 mAh cell, which was amplified using a voltage booster circuit through LD38050 PU IC, with an amplifying voltage from 0 to 56 V. This variable defines the intensity of the electric pulses. The voltage applied to the microcontroller was regulated to 3.3 V using the voltage regulator circuit—LM25775 IC. The current flowing to the body through the electrodes was controlled using a current limiter circuit developed using LM317T IC. It was used to limit the flow of current into the body within a permissible range. The printed circuit board was prepared based on the designed circuit. The assembly was placed in a customized 3D printed case. The components of the module are shown in Figure 8. A schematic representation of TENS module and kneecap on the knee are shown in Figure 9a and the kneecap worn on the knee and TENS module held in hand are shown in Figure 9b.

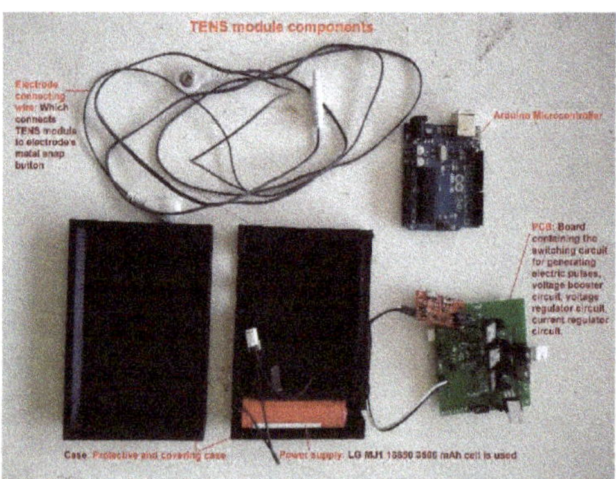

Figure 8. TENS module components.

The developed TENS module was compared with two other brands of TENS modules, marked as brands X and Y for confidentiality. The user feedback and average rating were used for the evaluation.

Output electric pulses were passed through the electrodes to the human body. Three different waveforms (Figure 10) were selected. Different burst waveforms were created, which could be modified for the attainment of different comforts and feels, depending on the user. The waveform specifications are presented in Table 3.

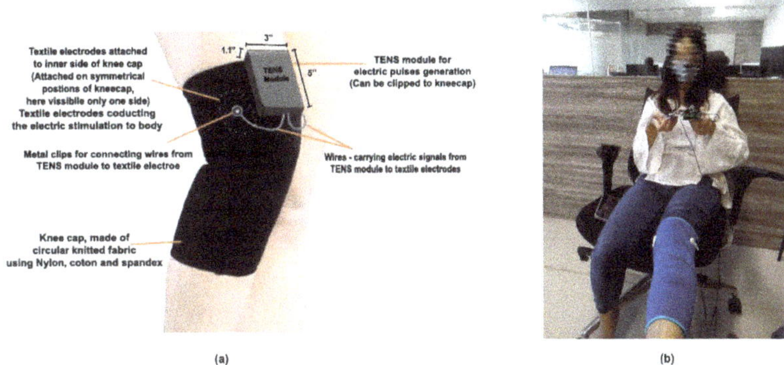

Figure 9. Schematic representation of the subject wearing the TENS module. (**a**) Schematic diagram of TENS module placed on kneecap, which is worn on knee. (**b**) Subject wearing kneecap and holding the TENS module in hand.

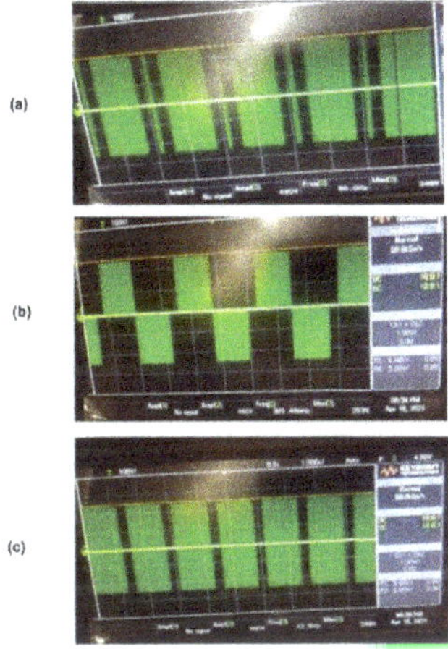

Figure 10. The images in (**a–c**) represent different symmetric waveforms selected as output electric waveforms, which can be selected or varied by the modification of microcontroller programming.

Table 3. Waveform specifications.

Waveform	Frequency	Description
a	100 Hz	Normal burst of 1200 ms long pulses with a smaller burst of 120 ms long; 330 ms gap
b	100 Hz	Positive and negative alternative bursts with 330 ms gap on each cycle; pulse duration is 1200 ms
c	60 Hz	Burst (60 Hz burst with 300 ms gap); 1200 ms long pulses

3. Results and Discussion

3.1. Testing

The smart kneecap with electric stimulation using the textile electrode was successfully developed, based on this study's objective. The developed circuit was able to generate a perfect symmetric waveform using the switching circuit and microcontroller. Three burst waveforms of varying frequencies and pulse width (Figure 10) were generated and studied.

Resistance of the textile electrode was found to be in the range of 0.4 to 2.1 ohm. The resistance value for the case of TENS with the hydrogel and carbon rubber as an electrode was found to be in mega ohm range. Therefore, the textile electrode exhibited higher efficiency for the TENS module. Notably, it was observed that the textile electrode required a voltage range of 20–25 V for the stimulation, which was in the range of 50–60 V in the case of hydrogel or carbon rubber electrode. Figure 11 presents the graph with average values of electric intensities for textile and hydrogel electrodes during the stimulation steps. Textile electrodes were found to be more suitable as an electrode in the TENS module.

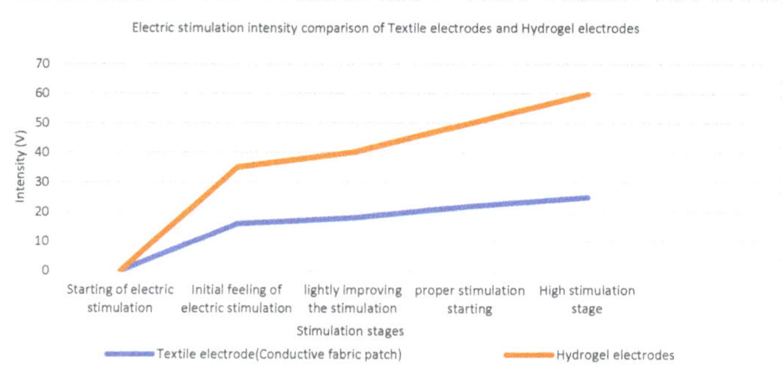

Figure 11. Intensity comparison of textile and hydrogel electrodes.

3.2. Feedback from Subjects

The voltages required for textile, hydrogel, and carbon rubber electrodes were analyzed. Typically, this is stated as the intensity of input power of electric pulses, which is required for the user to feel an effective stimulation. The input voltage requirements were compared among the textile electrodes and other conventional electrodes. The underlying challenge was the effect of individual user's skin impedance and their sense to electric stimulations. Therefore, a comparison of electrodes was conducted with respect to the voltage range, which is effective and comfortable to each individual. As a result, a comparison of all three types of electrodes (textile, hydrogel, and carbon rubber), was conducted for the individual user for the input voltage specific to each of them.

Feedbacks were collected from 20 users (human subjects). They were advised to sense the stimulation for a specific time of exposure for all the three TENS modules without conveying the brands/category. The feedback was collected on the basis of comfort, skin irritation, and sensitivity to electric stimulation. The developed module received better acceptability with the highest average feedback rating of 7.77 (out of 10). Furthermore, the four textile electrodes, prepared by the conductive fabrics from brands P, Q, R, and S, were also tested using subjective feedback. Brands P and Q obtained an average rating of 7.5, while brands R and S received an average rating of 3.5 and 2.5, respectively (out of 10).

4. Conclusions

In this research project, a smart kneecap integrated with a textile electrode-based TENS module was successfully developed. The electrodes were attached to the kneecap, which resulted in a wearable device. This study affirmed that a wide range of applications of conductive fabrics and textile electrodes are wellness-related wearable products. The developed TENS module was portable in handheld form or attached to the kneecap. In addition, it can operate at lower voltage and power, thus enabling a longer battery life when compared with hydrogel and carbon rubber electrodes. Symmetrical, stable waveforms, as well as smooth textile electrodes can lead to a better comfort level during the electric stimulation process on users (subjects).

As a future scope of this research project, a more compact structure could be attempted for the development of TENS module using a self-powered module integrated with a smaller form factor of microcontrollers. In response to this, the microcontrollers could be connected via Bluetooth technology through a mobile application for an interactive and remote operation.

Author Contributions: This research article was conducted by J.V.R., as part of his research project for the course Master of Fashion technology from National Institute of Fashion Technology, New Delhi. The research was conducted under the supervision of P.J., National Institute of Fashion Technology, New Delhi and R.T., National Institute of Fashion Technology Mumbai. All authors have read and agreed to the published version of the manuscript.

Funding: Financial support for this research was provided by Fupro Innovation Pvt. Ltd.

Institutional Review Board Statement: Not applicable.

Informed Consent Statement: Not applicable.

Acknowledgments: The authors acknowledge the opportunity provided to carry out the research work at Fupro Innovation Pvt. Ltd., an Indian start-up under the incubation of Tynor Orthotics Pvt. Ltd.

Conflicts of Interest: The authors declare no conflict of interest.

References

1. Jones, I.; Johnson, M.I. Transcutaneous electrical nerve stimulation. *Contin. Educ. Anaesth. Crit. Care Pain* **2009**, *9*, 130–135. [CrossRef]
2. Watson, T. Transcutaneous Electrical Nerve Stimulation (TENS). Physiopedia. Available online: https://www.physio-pedia.com/Transcutaneous_Electrical_Nerve_Stimulation_(TENS) (accessed on 4 March 2022).
3. Singh, J. *Textbook of Electrotherapy*, 3rd ed.; Jaypee Brothers Medical Publishers: New Delhi, India, 2012.
4. Vance, C.G.; Dailey, D.L.; Rakel, B.A.; Sluka, K.A. Using TENS for pain control: The state of the evidence. *Pain Manag.* **2014**, *4*, 197–209. [CrossRef] [PubMed]
5. Zhou, H.; Lu, Y.; Chen, W.; Wu, Z.; Zou, H.; Krundel, L.; Li, G. Stimulating the Comfort of Textile Electrodes in Wearable Neuromuscular Electrical Stimulation. *Sensors* **2015**, *15*, 17241–17257. [CrossRef] [PubMed]
6. Erdem, D.; Yesilpinar, S.; Senol, Y. Design of TENS electrodes using conductive yarn. *Int. J. Cloth. Sci. Technol.* **2016**, *28*, 311–318. [CrossRef]
7. Yapici, M.K.; Alkhidir, T.; Samad, Y.A.; Liao, K. Graphene-clad textile electrodes for electrocardiogram monitoring. *Sens. Actuators. B Chem.* **2015**, *221*, 1469–1474. [CrossRef]
8. Márquez Ruiz, J.C. On the Feasibility of Using Textile Electrodes for Electrical Bioimpedance Measurements. Ph.D. Thesis, KTH–Royal Institute of Technology, Stockholm, Sweden, 2011.

engineering proceedings

MDPI

Proceeding Paper

Sensing of Body Movement by Stretchable Triboelectric Embroidery Aimed at Healthcare and Sports Activity Monitoring †

Hasan Riaz Tahir *, Benny Malengier, Granch Berhe Tseghai and Lieva Van Langenhove

Department of Materials, Textiles and Chemical Engineering, Ghent University, 9000 Ghent, Belgium; benny.malengier@ugent.be (B.M.); granchberhe.tseghai@ugent.be (G.B.T.); lieva.vanlangenhove@ugent.be (L.V.L.)
* Correspondence: hasan.tahir@ugent.be; Tel.: +32-47-293-6221
† Presented at the 3rd International Conference on the Challenges, Opportunities, Innovations and Applications in Electronic Textiles (E-Textiles 2021), Manchester, UK, 3–4 November 2021.

Abstract: In this work, we introduced an embroidery-based stretchable (up to 60–70%) triboelectric nano-generator that could be attached to different parts of the human body such as fingers, knee, elbow, back, or shoulders, to sense the body movement. It can be used as activity recognition for health care and sport activity monitoring. The sensor was composed of different yarns embroidered on a stretchable conductive substrate, allowing it to sense diverse mechanical deformation of different body parts. Different stitching styles, patterns, stitch lengths, and shapes have been selected to cater to the unidirectional, bidirectional, and multidirectional force and obtain the maximum movement flexibility. In order to do embroidery on a stretchable substrate, a non-stretchable water-soluble second substrate has been added before embroidering, and is afterwards removed by application of steam. A sample of 1.5×6 cm^2 was used for sensing finger movement and generated a peak to peak voltage of 274.5 mV. The amount of generated voltage depended upon the application area on the body and its deformation, thread type, stitch type, stitch length, and shape of embroidery. A stitch length of more than 2 mm with a line density of 1 line per mm resulted in a stretchable sample. The state of the art of the developed sensors is their low price, flexibility, and low weight. They are all obtained with commercially available embroidery yarns and commercially available technology for their development.

Keywords: stretchable sensor; embroidery; triboelectric yarns; activity monitoring; flexible; mechanical deformation

Citation: Tahir, H.R.; Malengier, B.; Tseghai, G.B.; Van Langenhove, L. Sensing of Body Movement by Stretchable Triboelectric Embroidery Aimed at Healthcare and Sports Activity Monitoring. *Eng. Proc.* **2022**, *15*, 4. https://doi.org/10.3390/engproc2022015004

Academic Editors: Russel Torah, Steve Beeby and Kai Yang

Published: 9 March 2022

Publisher's Note: MDPI stays neutral with regard to jurisdictional claims in published maps and institutional affiliations.

1. Introduction

Electronic textiles (e-textiles) have received significant attention because of their remarkable application in wearable electronics [1–4]. In recent years, the interest in stretchable and deformable wearable electronics has grown due to their potential application in several fields, including sports [5] and healthcare [6,7]. In addition, activity recognition is a crucial parameter that plays a vital role in healthcare monitoring [8] and can be achieved with e-textiles. However, all these require power, and the friction between two different materials can generate an electrostatic charge that can be harvested by a conductive substrate [9,10].

Stretchable sensors are usually made with stretchable carbon-based materials or textile structures. Different efforts have been made to develop stretchable and wearable sensors by different processes, including highly stretchable single electrodes triboelectric nanogenerator (TENGs) based on conductive nanowires and an ultrathin dual-mode patch acting as a self-powered sensor [11–13]. However, these are high-cost solutions. In this research work, we developed body movement sensors for healthcare and sports activity monitoring from commercially available embroidery yarns from Madeira, Germany, and with standard semi-professional embroidery machines.

2. Methodology and Results

To develop these body activity monitoring sensors, the following commercially available Madeira embroidery yarns were used.

1. Madeira 100% Viscose (VIS);
2. Madeira PES-100% Polyester Embroidery Thread (PES);
3. 60% Polybutylene Terephthalate (PBT)/40% Polypropylene (PP) Madeira;
4. Polyneon 100% Polyester Madeira (PN PES);
5. 100% Aramid Madeira (AR).

In order to cater to the different kinds of forces (unidirectional, bidirectional, multi-all directional) which are present upon application at different locations on the body, a selection of stitch types and shapes has been made to obtain a maximum peak to peak voltage while retaining the flexibility of the structure. The stitch length of 2 mm, with a stitch line density of 1 line per mm, has been selected to give flexibility to the embroidery structure.

Figure 1 shows the different stitch types and shape selection according to the application area of the body so that the generator allows maximum flexibility of movement. The most critical and challenging part was to perform the embroidery on a stretchable conductive substrate. As a substrate, Shieldex ® Technick-Tex P130+B (Statex Produktions- und Vertriebs GmbH, Bremen, Germany), a two-way stretchable knitted fabric consisting of 22% elastomer and 78% polyamide, was chosen. This conductive fabric was purchased from Statex Produktions- und Vertriebs GmbH Kleiner Ort 9-11, 28357 Bremen, Germany. The fabric's GSM (Gram per square meter) is 132 g/m^2 with a thickness of 0.55 mm.

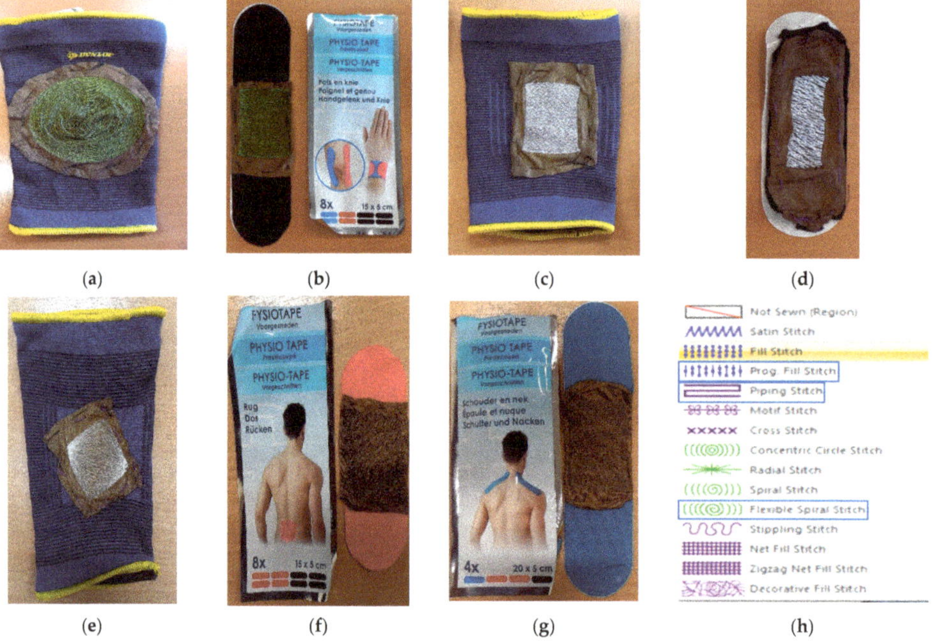

(a) (b) (c) (d)

(e) (f) (g) (h)

Figure 1. This figure shows the different body movement sensors with compression sleeves, physio tape and selection of four stitch types that give the maximum flexibility of movement (**a**) knee sensor with knee bandage; (**b**) knee sensor with physio tape; (**c**) elbow sensor with elbow bandage (**d**) finger movement sensor (**e**) elbow sensor with piping stitch and rhombus shape for maximum flexibility of movement (**f**) back movement sensor with physio tape (**g**) shoulder movement sensor with physio tape (**h**) selection of stitch types (Fill, Prog. Fill, Piping, and Flexible spiral stitch).

It is not possible to embroider directly on a conductive stretchable substrate. In order to solve this problem, a water-soluble non-stretchable sheet was attached so that the stretchable substrate does not go into the machine parts during the embroidery process. After completion, the water-soluble non-stretchable substrate was removed by applying steam, and the sample became stretchable again. In order to have body activity monitoring, the prototypes and different waveforms are shown in Figures 1 and 2. They were applied to the moving parts of the body through compression sleeves or physio tapes.

The characterization is performed by directly attaching the sensors to an oscilloscope. The body activity monitoring sensors generate a voltage waveform captured with an oscilloscope RIGOL DS2102A as summarized in Table 1. Table 1 shows the stitch type, shape, application area, body part, and the resulting peak to peak voltage under movement. The maximum generated peak to peak voltage was 1073 mV obtained with polyester yarns and flexible spiral stitch at a stitch length of 2 mm with physio tape and application on the knee joint. The generated voltage is the consequence of the inter friction of different yarns and the friction of the conductive substrate with the applied material during the stretching and/or bending movement.

State of the art: The state of the art of the presented triboelectric nano-generators is their low price, flexibility, and low weight, all obtained with commercially available embroidery yarns and commercially available technology for their development.

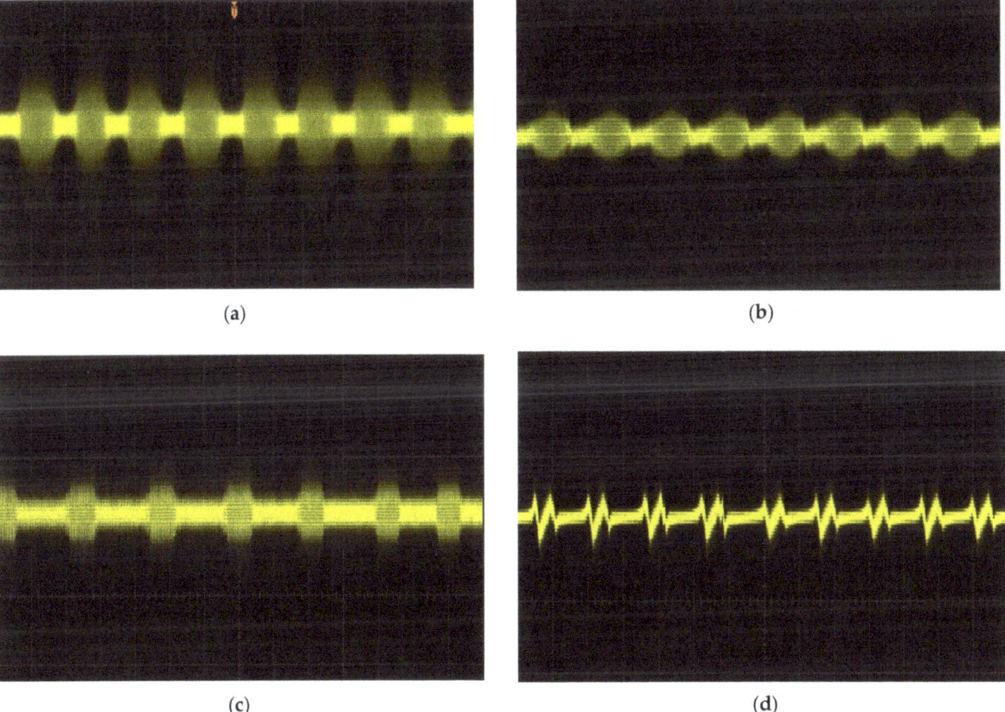

(a)

(b)

(c)

(d)

Figure 2. The effect on the waveform for different application area and stich type: (**a**) waveform for knee movement developed with Polyneon 100% Polyester Madeira (PN PES) using fill stitch; (**b**) waveform for elbow movement developed with 60% PBT/40% PP Madera using pipping stitch; (**c**) waveform for finger movement developed with viscose and polyester Madera using Prog. Fill stitch; (**d**) waveform for knee movement developed with polyester Madeira yarn and through compression sleeves.

Table 1. Stretchable sensors stitch type, shape, application area, and the triboelectric results.

Yarn Type	Stitch Type	Shape	Size (cm^2)	Application by	Application Part	Peak to Peak mV	Current Density μA cm^{-2}	Power Density nW cm^{-2}
VIS/PES	Prog. Fill	Rectangle	1.5 × 6	Physio Tape	Finger	274.5	0.0305	8.37
VIS	Piping	Rhombus	6 × 6	Bandage	Elbow	126.0	0.0035	0.441
		Rectangle	4.5 × 6	Bandage	Knee	178.0	0.0066	1.17
PES	Flexible Spiral	Square	4 × 4	Bandage	Elbow	219.2	0.0137	3.00
		Round	9.5	Bandage	Knee	1073	0.0151	16.25
60% PBT/40% PP	Piping	Rectangle	7 × 4	Physio Tape	Knee	223.8	0.0080	1.79
PN PES 100%	Fill	Rectangle	4 × 6	Physio Tape	shoulder	105.9	0.0044	0.47
AR	Piping-longer stitch length	Rectangle	4 × 5	Physio Tape	Back	111.2	0.0056	0.62

3. Conclusions

Embroidery on the stretchable conductive substrate was challenging but possible with a soluble substrate. After embroidering, steam is applied to disintegrate the non-stretchable part. The stitch length and stitch line density play an essential part in obtaining a stretchable structure. The stitch length of 2 mm and line density of 1 line per mm has been selected, with particular patterns, to obtain the desired stretch. At lower stitch length and stitch density, it is not possible to have this stretch. Additionally, the stitch type of the pattern is essential. The following four stitch types were selected: Fill, Prog. Fill, Piping, Flexible spiral. These stitch types are essential to get the stretchable structure with the selected stitch length.

The resulting waveform under movement depended mainly upon the body's application area and the stitch type, stitch length, and shape of embroidery. The state of the art is their low price, flexibility, and lightweight, with commercially available embroidery yarns and standard embroidery equipment for their development. The most important application is for activity recognition in health care and sports activity monitoring.

Author Contributions: H.R.T. conceived the idea, conducted the experiment, and wrote the paper; G.B.T. and B.M. helped in experimenting and edited the manuscript; L.V.L. supervised and administrated the project. All authors have read and agreed to the published version of the manuscript.

Funding: The research was funded by ICT-Tex project EU project (Nr. 612248-EPP-1-2019-1-BGEPPKA2-KA) and HEC (Higher Education Commission), Pakistan: HRDI-UESTP Scholarship Project.

Institutional Review Board Statement: Not applicable.

Informed Consent Statement: Not applicable.

Acknowledgments: The authors would also like to thank the Ingegno Maker Space, Drongen, Belgium for the use of their digital embroidery machine and equipment in the creation of the flexible sensors.

Conflicts of Interest: The authors declare no conflict of interest.

References

1. Ling, W.; Liew, G.; Li, Y.; Hao, Y.; Pan, H.; Wang, H.; Ning, B.; Xu, H.; Huang, X. Materials and Techniques for Implantable Nutrient Sensing Using Flexible Sensors Integrated with Metal-Organic Frameworks. *Adv. Mater.* **2018**, *30*, e1800917. [CrossRef] [PubMed]
2. Du, D.; Li, P.; Ouyang, J. Graphene coated nonwoven fabrics as wearable sensors. *J. Mater. Chem. C* **2016**, *4*, 3224–3230. [CrossRef]

3. Heo, J.S.; Eom, J.; Kim, Y.H.; Park, S.K. Recent progress of textile-based wearable electronics: A comprehensive review of materials, devices, and applications. *Small* **2018**, *14*, 1703034. [CrossRef] [PubMed]
4. Xue, Q.; Sun, J.; Huang, Y.; Zhu, M.; Pei, Z.; Li, H.; Wang, Y.; Li, N.; Zhang, H.; Zhi, C. Recent Progress on Flexible and Wearable Supercapacitors. *Small* **2017**, *13*. [CrossRef] [PubMed]
5. Zhu, M.; Shi, Q.; He, T.; Yi, Z.; Ma, Y.; Yang, B.; Chen, T.; Lee, C. Self-Powered and Self-Functional Cotton Sock Using Piezoelectric and Triboelectric Hybrid Mechanism for Healthcare and Sports Monitoring. *ACS Nano* **2019**, *13*, 1940–1952. [CrossRef] [PubMed]
6. Meng, K.; Zhao, S.; Zhou, Y.; Wu, Y.; Zhang, S.; He, Q.; Wang, X.; Zhou, Z.; Fan, W.; Tan, X.; et al. A Wireless Textile-Based Sensor System for Self-Powered Personalized Health Care. *Matter* **2020**, *2*, 896–907. [CrossRef]
7. Jao, Y.-T.; Yang, P.-K.; Chiu, C.-M.; Lin, Y.-J.; Chen, S.-W.; Choi, D.; Lin, Z.-H. A textile-based triboelectric nanogenerator with humidity-resistant output characteristic and its applications in self-powered healthcare sensors. *Nano Energy* **2018**, *50*, 513–520. [CrossRef]
8. Wang, Y.; Cang, S.; Yu, H. A survey on wearable sensor modality centred human activity recognition in health care. *Expert Syst. Appl.* **2019**, *137*, 167–190. [CrossRef]
9. Fan, F.-R.; Tian, Z.-Q.; Wang, Z.L. Flexible triboelectric generator. *Nano Energy* **2012**, *1*, 328–334. [CrossRef]
10. Liu, S.; Zheng, W.; Yang, B.; Tao, X. Triboelectric charge density of porous and deformable fabrics made from polymer fibers. *Nano Energy* **2018**, *53*, 383–390. [CrossRef]
11. Cui, N.; Gu, L.; Lei, Y.; Liu, J.; Qin, Y.; Ma, X.-H.; Hao, Y.; Wang, Z.L. Dynamic Behavior of the Triboelectric Charges and Structural Optimization of the Friction Layer for a Triboelectric Nanogenerator. *ACS Nano* **2016**, *10*, 6131–6138. [CrossRef] [PubMed]
12. Pu, X.; Liu, M.; Chen, X.; Sun, J.; Du, C.; Zhang, Y.; Zhai, J.; Hu, W.; Wang, Z.L. Ultrastretchable, transparent triboelectric nanogenerator as electronic skin for biomechanical energy harvesting and tactile sensing. *Sci. Adv.* **2017**, *3*, e1700015. [CrossRef] [PubMed]
13. Chen, X.; Song, Y.; Chen, H.; Zhang, J.; Zhang, H. An ultrathin stretchable triboelectric nanogenerator with coplanar electrode for energy harvesting and gesture sensing. *J. Mater. Chem. A* **2017**, *5*, 12361–12368. [CrossRef]

engineering proceedings

MDPI

Proceeding Paper

Knitted Graphene Supercapacitor and Pressure-Sensing Fabric †

Yi Zhou [1,*], Chunyan Zhang [2], Connor Myant [1] and Rebecca Stewart [1]

1   Dyson School of Design Engineering, Imperial College London, London SW7 2DB, UK;
    connor.myant@imperial.ac.uk (C.M.); r.stewart@imperial.ac.uk (R.S.)
2   College of Civil Aviation, Shenyang Aerospace University, Shenyang 110136, China;
    192113065221@email.sau.edu.cn
*   Correspondence: y.zhou20@imperial.ac.uk
†   Presented at the 3rd International Conference on the Challenges, Opportunities, Innovations and Applications
    in Electronic Textiles (E-Textiles 2021), Manchester, UK, 3–4 November 2021.

Abstract: This research utilizes a simple and effective dip coating/ultrasonication method to prepare porous graphene-coated sensing fabrics made with commercially produced acrylic/spandex yarn with multifunctional performance. We examine the electrochemical performance of graphene-coated fabrics and explore their potential in applications involving pressure sensors. The results show that our graphene-coated fabric demonstrates a maximum specific capacitance value of 17.4 F/g. When applied as a pressure sensor, the capacitance change rate of our sensor increases linearly with the increase in pressure applied to the fabrics. Our sensor also shows a fast response in a pressure loading–unloading test, which indicates an outstanding sensing property and shows promising capabilities as a supercapacitor.

Keywords: e-textiles; graphene; supercapacitor; pressure sensor

Citation: Zhou, Y.; Zhang, C.; Myant, C.; Stewart, R. Knitted Graphene Supercapacitor and Pressure-Sensing Fabric. *Eng. Proc.* **2022**, *15*, 5. https://doi.org/10.3390/engproc2022015005

Academic Editors: Russel Torah, Steve Beeby and Kai Yang

Published: 9 March 2022

Publisher's Note: MDPI stays neutral with regard to jurisdictional claims in published maps and institutional affiliations.

1. Introduction

Stretchable and wearable electronics have attracted extensive attention from both academia and industry due to their unique flexibility, deformability, and portability [1]. In addition, development of energy storage devices with good electrochemical performance and flexibility to meet the energy requirements of flexible wearable electronics has become one of the major research directions in recent years due to the increasing prevalence of wearable technologies [2].

Among various conductive materials, graphene has become an ideal material for flexible electrodes because of their extremely high specific surface area, mechanical stiffness, and outstanding electrical conductivity [3,4]. The combination of textiles and graphene maintains the original physical properties of the fabric and provides a support for active materials while providing a stable conductive surface. Therefore, graphene-coated fabrics are considered an ideal material for electrodes [5].

In this study, we design and knit graphene-coated sensing fabrics with acrylic/spandex yarns through a cost-effective dip-coating method. The microstructure, and electrochemical and sensing properties of the porous graphene-coated sensing fabrics are examined to investigate their capabilities as a supercapacitor and pressure sensor.

2. Materials and Methods

2.1. Graphene-Coated Knitted Fabrics

We designed knitted fabrics with a full needle knitting structure, using a gauge 10 Dubied knitting machine, 25 rows in height, and with 45 needles on a double bed to knit the elastic yarn (90% acrylic and 10% spandex) into the same dimensions (20 mm × 60 mm × 1.5 mm).

The prepared knitted fabrics then were then fabricated into sensing fabrics using the dip-coating method, shown in Figure 1. The fabrics were purified by soaking them in ethanol and deionized water washings. After drying, the fabrics were dipped into a stable 1.2 wt.% graphene/acetone suspension under sonication for 30 min and then dried at room temperature. After three cycles of the coating process, the fabrics were washed with deionized water and then dried in a furnace at 40 °C.

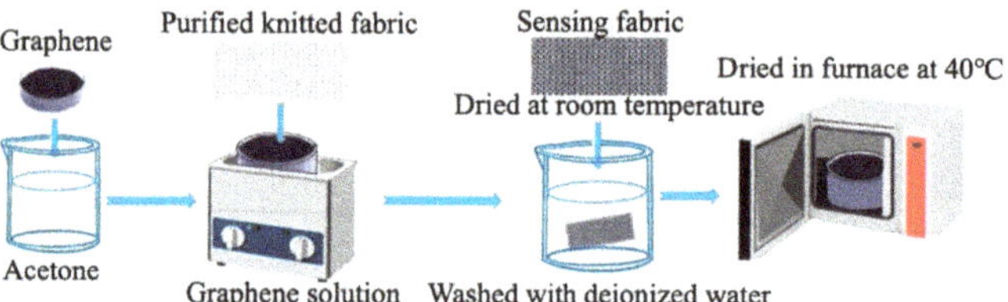

Figure 1. Fabrication process of graphene-coated sensing fabrics.

2.2. Electrochemical and Sensing Property Tests

The electrochemical properties of our graphene-coated fabrics were examined by testing cyclic voltammetry (*CV*), galvanostatic charge/discharge (*GCD*), cyclic stability measurements, and electrochemical impedance spectroscopy (*EIS*). A three-electrode supercapacitor device was set up, as illustrated in Figure 2, using 1 M Na2SO4 solution as the electrolyte, a Pt sheet as counter electrode, a Ag/AgCl sheet as reference electrode and our graphene-coated fabric as the working electrode. The *CV* curves were generated at a scan rate from 20 mV/s to 100 mV/s. The specific capacitance (*C*) was calculated from the area in the *CV* curves using the following equation:

$$C = \frac{\int I \, dv}{v \, \triangle V m} \tag{1}$$

Using the same three-electrode setup, the *GCD* measurements were tested at current densities of 0.5, 1, 1.5 and 2 A/g. The specific capacitance value with the current densities in the *GCD* curves was calculated using the following equation:

$$C = \frac{I \cdot \triangle t}{m \cdot \triangle v} \tag{2}$$

where *I* = current (A), *v* = scan rate (V·s−1), *V* = working potential, *m* = weight of graphene in the fabrics (g), and Δ*t* = discharging time.

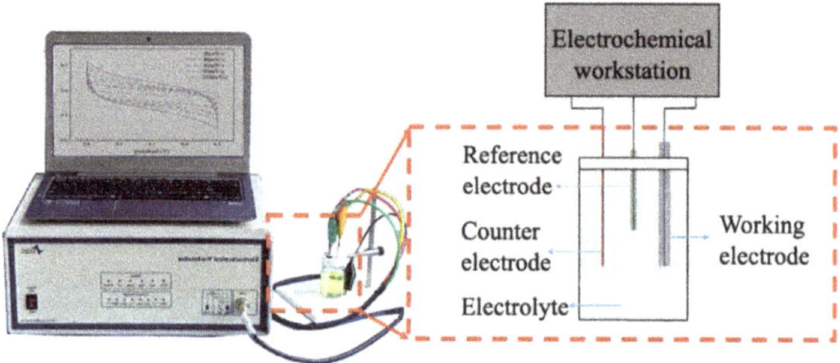

Figure 2. Schematic diagram of three-electrode supercapacitor system connected to electrochemical workstation.

The cyclic stability test was conducted at a current density of 2 A/g, charging and discharging 10,000 times.

The Nyquist curve was produced using the *EIS* and it measured the impedance through a decreasing AC frequency. The ion diffusion resistance of our graphene-coated fabric was calculated using Zview software.

For the sensing property test, various pressures were applied to the sensing fabric, and the electrochemical station was applied to measure the capacitance of the fabric. Then, the response time of our graphene-coated fabric was tested under external loading and unloading with a pressure of 24.5 kpa. The change rate in the capacitance was calculated using the ratio of change in capacitance ($\triangle C$) to the original capacitance (C_0).

3. Results and Discussion

3.1. Morphology of Graphene-Coated Fabrics

A scanning electron microscope (SEM) was applied to analyze the surface of our graphene-coated fabrics at different magnifications with the operating voltage of 5 kV.

Figure 3a is an overview of the knitted fabrics, showing porous structures that allows deposition of charged ions. It can be seen from Figure 3b that acrylic fibers form a uniform interweaved structure with multiple graphene nanoplatelets (GNPs) dispersed on the fiber surface and between fiber gaps. Figure 3c shows a typical crinkled structure of GNPs coating the textile fibers. Overall, the porous structure of our knitted fabrics can provide sufficient surface are for energy storage, which indicates the potential for incorporating such materials into a supercapacitor.

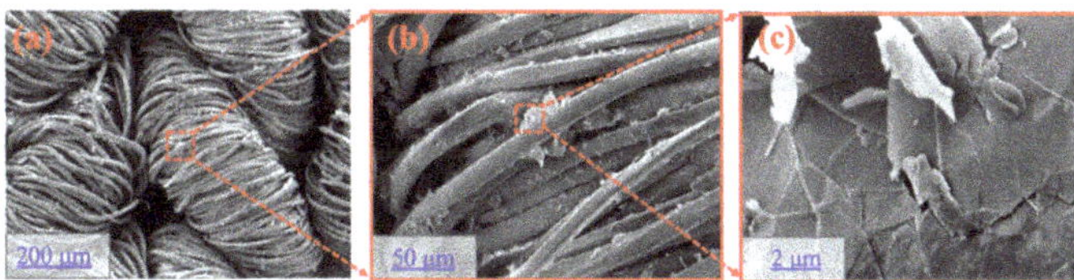

Figure 3. SEM images of graphene-coated knitted fabrics at (**a**) 200 μm, (**b**) 50 μm and (**c**) 2 μm.

3.2. Electrochemical and Sensing Performance of Graphene-Coated Fabrics

In the electrochemical performance tests, a *CV* test was first conducted, as shown in Figure 4a. All the *CV* curves at different scanning rates show a similar rectangular shape and demonstrate a maximum specific capacitance of 17.4 F/g at a scan rate of 20 mv/s. Figure 4b shows the *GCD* curves of graphene-coated fabrics. When the current density increases from 0.5 A/g to 2 A/g, the corresponding specific capacitance decrease from 16.7 F/g to 7.5 F/g. The *GCD* curves at different current densities all show similar trends with a triangular curve, indicating ideal capacitor performance. Figure 4c reflects the cyclic stability of the graphene-coated fabrics. After 10,000 cycles of charging and discharging, the capacitance of our sensing fabric remained above 90%. Figure 4d demonstrates the Nyquist curve of the graphene-coated fabrics. The almost negligible semicircle diameter for the graphene-coated fabrics reveals that the porous structure of the knitted fabrics facilitates a good contact between the electrode and electrolyte. The ion diffusion resistance of the graphene-coated fabric was calculated to be 20 Ω, demonstrating a good ion transport ability.

We further applied our graphene-coated fabric as a pressure sensor to test its sensing performance. Figure 5a shows that the capacitance change rate of our sensor increases linearly with the increase of pressure applied to the fabric. When the pressure was increased to 24.5 kPa, the capacitance change rate of the graphene-coated fabric reached its maximum value of 28%. Figure 5b demonstrates that our sensor had a fast response time with a recovery time of 0.6 s and 0.4 s under external loading and unloading, respectively. Our graphene-coated fabrics showed satisfactory performance as supercapacitors and as flexible sensors, indicating their great potential in further practical applications.

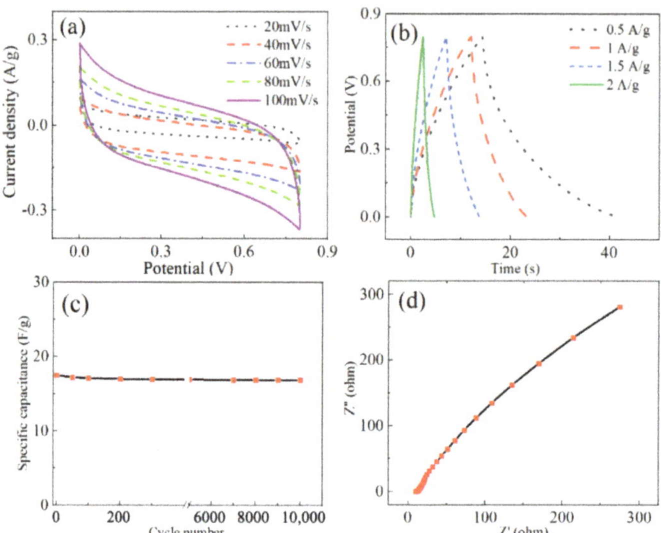

Figure 4. (**a**) CV curves of graphene-coated knitted fabrics at different scanning rates; (**b**) GCD curves of graphene-coated knitted fabrics; (**c**) cyclic stability test of graphene-coated knitted fabrics; (**d**) Nyquist diagram of graphene-coated knitted fabrics.

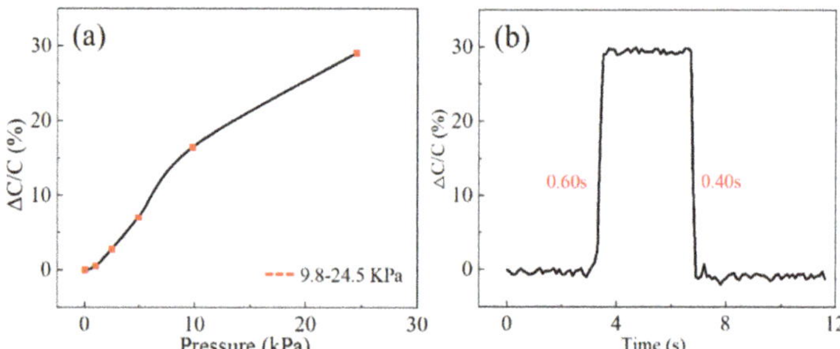

Figure 5. (**a**) Rate of change in capacitance as pressure applied within 0–25 kPa; (**b**) response and recovery time test under loading and unloading pressure.

4. Conclusions

This paper presents a graphene-coated sensing fabric that can be applied as a supercapacitor or pressure sensor by using a simple dip coating and ultrasonication method. The sensing fabric based on double bed weft-knitted fabric using acrylic/spandex yarns exhibits the ability to be charged/discharged. When applied as an electrode, our graphene-coated fabric demonstrates a maximum specific capacitance value of 17.4 F/g. Additionally, the graphene-coated fabric can be applied as a pressure sensor, the maximum capacitance change rate of our sensing fabric is up to 28%, and the compression response and recovery time are 0.6 s and 0.4 s. Since this research only explored the potential for graphene-coated fabric being applied as a supercapacitor or pressure sensor, further work will consider the stability and effect of wear due to washing. In addition, we will explore the reliability and longevity of our sensing fabric in a device or product.

Author Contributions: Conceptualization, Y.Z. and C.Z.; methodology, C.Z.; formal analysis, Y.Z.; investigation, Y.Z.; resources, Y.Z. and C.Z.; writing—original draft preparation, Y.Z.; writing—review and editing, R.S.; visualization, Y.Z.; supervision, C.M. and R.S.; project administration, C.M. and R.S. All authors have read and agreed to the published version of the manuscript.

Funding: This research received no external funding.

Institutional Review Board Statement: This study did not require ethical approval.

Informed Consent Statement: Not applicable.

Data Availability Statement: Not applicable.

Acknowledgments: The author would like to thank researchers from Shenyang Aerospace University for their kind support. In particular, we thank Qingshi Meng and Yin Yu for their advice and help with sensor fabrication and data processing.

Conflicts of Interest: The authors declare no conflict of interest.

References

1. Shi, Q.; Lee, C. Self-Powered Bio-Inspired Spider-Net-Coding Interface Using Single-Electrode Triboelectric Nanogenerator. *Adv. Sci.* **2019**, *6*, 1900617. [CrossRef] [PubMed]
2. Liu, Z.; Mo, F.; Li, H.; Zhu, M.; Wang, Z.; Liang, G.; Zhi, C. Advances in Flexible and Wearable Energy-Storage Textiles. *Small Methods* **2018**, *2*, 1800124. [CrossRef]
3. Meng, Q.; Yu, Y.; Tian, J.; Yang, Z.; Guo, S.; Cai, R.; Han, S.; Liu, T.; Ma, J. Multifunctional, durable and highly conductive graphene/sponge nanocomposites. *Nanotechnology* **2020**, *31*, 465502. [CrossRef] [PubMed]
4. Cui, X.; Zhang, C.; Araby, S.; Cai, R.; Kalimuldina, G.; Yang, Z.; Meng, Q. Multifunctional, flexible and mechanically resilient porous polyurea/graphene composite film. *J. Ind. Eng. Chem.* **2022**, *105*, 549–562. [CrossRef]
5. Li, Y.; Zhang, Y.; Zhang, H.; Xing, T.; Chen, G. A facile approach to prepare a flexible sandwich-structured supercapacitor with rGO-coated cotton fabric as electrodes. *RSC Adv.* **2019**, *9*, 4180–4189. [CrossRef]

engineering proceedings

MDPI

Proceeding Paper

Ambulatory Monitoring Using Knitted 3D Helical Coils [†]

Kristel Fobelets * and Christoforos Panteli

Electrical and Electronic Engineering, Imperial College London, Exhibition Road, London SW7 2AZ, UK;
christoforos.panteli11@imperial.ac.uk
* Correspondence: k.fobelets@imperial.ac.uk
† Presented at the 3rd International Conference on the Challenges, Opportunities, Innovations and Applications
  in Electronic Textiles (E-Textiles 2021), Manchester, UK, 3–4 November 2021.

Abstract: We present a highly sensitive wearable angular position sensor to measure joint movement. The sensor is a 3D helical coil knitted in the sleeve of a garment by circularly knitting thin insulated metal wire and yarn simultaneously. The sensing mechanism is based on the variation of the mutual inductance between windings. A 167 µH change is measured for knee movement from fully stretched to completely bent. A double cross coupled FET pair transforms the low-Q coils into a high-Q system giving a maximum frequency variation of 145 kHz for knee bending.

Keywords: motion tracking; inductive sensing; wearable; knitting

1. Introduction

Wearable sensors will play an important role in future health and wellbeing support approaches [1]. E-garments give patients more autonomy while clinicians can still access and evaluate the data remotely, when required. These sensors must be easy to wear and not hinder movement; read-out electronics must be low-power and compact; and the system easy to use and not require expert sensor placement. Wearables are candidates for deployment in rehabilitation, e.g., remote supervision of stroke victims' exercises. They can be used in sports performance training and other applications where observations of quality and duration of movement are useful. Movement sensors come in different shapes and forms. Many are based on inertial sensors, including accelerometers, gyroscopes and magnetometers with small form factor and 3D movement registration possibilities [2]. Other sensor types are based on resistive [3] and inductive changes [4]. The advantage of these is their simplicity and direct relationship between measured parameter and angle of a joint. Inductive sensors have the added benefit for wearables that no additional layers are required for electrical isolation. We present an alternative implementation of an inductive sensor that is a knitted helical coil that forms part of a garment. These knitted coils are easily implemented at the elbow in a sleeve or at the knee in a long sock. The inductor is circularly knitted, using yarn and thin insulated metal wire, forming a 3D helical coil [5]. Changes in the geometry of the coil changes its self-inductance which corresponds to the movement and the angle of a limb. This approach makes these garments indistinguishable from normal knitted clothes. In this work we demonstrate the feasibility of this implementation as a wearable ambulatory monitoring system.

2. Materials and Methods

Different sleeves were weft knitted by hand with circular needles using yarn and thin insulated metal wire simultaneously (Figure 1a). Sleeves were made in different diameters consistent with wrist, elbow and knee dimensions. The influence of the number of rows with metal (number of turns N), the type of metal wire (solid Cu magnet wire ~250 µm and Cu/Ag Litz wire ~150 µm diameter), the yarn and needle size and the knit style (jersey or 1/1 rib) were also investigated.

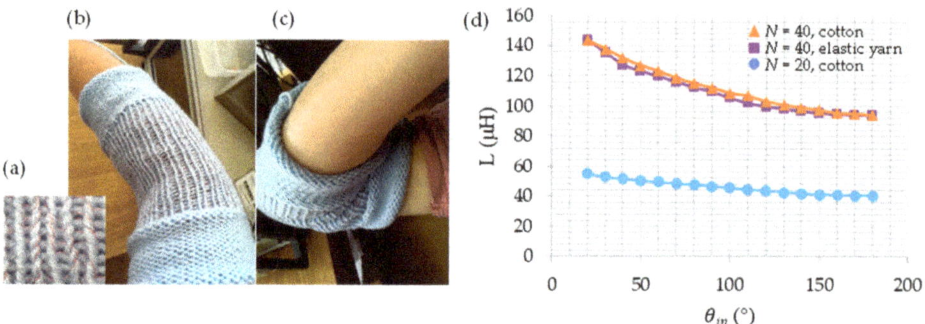

Figure 1. (**a**) Coil stitches with Cu wire (red). (**b**) Sleave at large internal angle giving minimum inductance. (**c**) Sleave at smallest internal angle giving maximum inductance. (**d**) Self-inductance of 1/1 rib elbow sleeves as a function of angle.

The measurements were performed using a protractor system as in [6] and a Wayne Kerr 6500B precision component analyzer to measure the self-inductance, L, and series resistance, R. A small 50 mV ac voltage was applied at 200 kHz, well below the self-resonance of the coil. Movement was measured on a healthy test subject (Figure 1b,c).

The self-inductance of the knitted coil can be approximated by summing the loop inductances L_{1loop} and the mutual inductance M_{1i} between different windings [7].

$$L = NL_{1loop} + 2\sum_{i=1}^{n-1}(N-i)M_{i1}(s \pm \Delta s) \tag{1}$$

where s is the distance between the windings and Δs is its variation due to bending of the coil. The expression for the mutual inductance is ($s_{1,2}$ are points on the coil):

$$M = \frac{\mu_0}{4\pi}\left(\oiint \frac{ds_1 \cdot ds_2}{|s_1 - s_2|}\right) \tag{2}$$

This shows that the mutual inductance strongly depends on distance between windings ($s_1 - s_2$). A reduction in distance increases M sharply. Thus, bending the knitted coil will bring the windings at the inner angle of the limb closer together resulting in a sharp increase in L. When the limb is stretched out ($\theta_{in} = 180°$), L is minimum (Figure 1b) and fully bent ($\theta_{in} \approx 0°$) L is maximum (Figure 1c). R does not change with bending.

3. Results and Discussion

The variation of L with inner angle θ_{in} for a sleave is given in Figure 1d. There is a steep response for small angles with the sensitivity decreasing for larger angles. This is as expected from Equations (1) and (2). The sensitivity decreases for lower N but its linearity increases. For a smaller region of $20° < \theta_{in} < 120°$, the variation of L with angle is nearly linear, given by the correlation coefficient R2. For $N = 40$ in 1/1 rib, Cu magnet wire and 3 mm needles, the wrist (not shown) sensitivity is 75 nH/° with $R^2 > 0.98$, the elbow 353 nH/° with $R^2 > 0.97$ and the knee (not shown) ~551 nH/°. The sensitivity increases with increasing sleeve diameter. Using elastic yarn did not change these findings (Figure 1d). Solid Cu magnet wire showed higher sensitivity than Cu/Ag Litz wire: 75 nH/° vs. 54 nH/°. The sensitivity of the knit was also slightly better than that of an inlaid coil that does not follow the horseshoe character of the stitches: 75 nH/° vs. 63 nH/°. The inlaid method has limited elasticity and is less suitable from a wearable's point of view.

Dynamic measurements were carried out on three different healthy subjects (two females, one male), moving the elbow through all angles. Figure 2a shows the variation of the normalized inductance L/L_{max} with movement as a function of time (time frame is ~6 s). There are two observations, whilst the absolute inductance for each subject is different because of different arm diameters, the normalized changes are nearly identical. At the start of the movement there are some memory effects in the knit (crinkling) that shift the minimum value of L. Once this initial shift has happened, no further changes occur until the crinkles are reset. The inset in Figure 2a gives the difference between changes in L when the sleave is worn on the elbow and when it is worn on the biceps whilst moving. The latter causes changes due to coil diameter changes in response to muscle movement. The change in inter-winding position is larger than that in diameter.

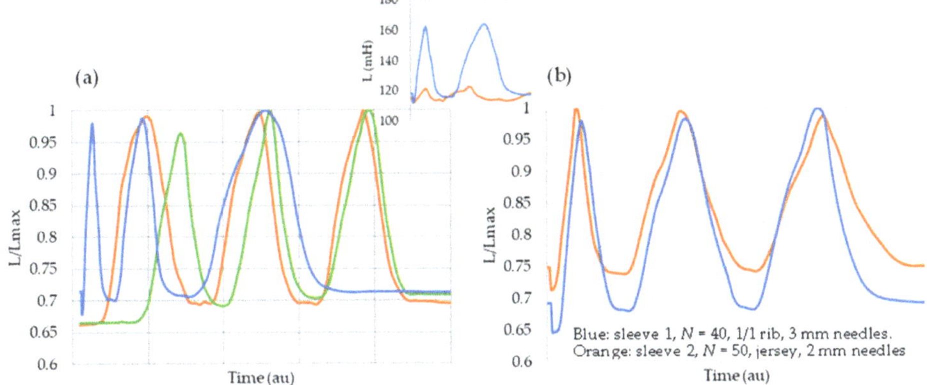

Figure 2. Dynamic normalized response L/L_{max}. (**a**) Elbow movement of three test subjects moving through the same angles at slightly different speeds. Inset: L in function of angle (blue) and diameter due to muscle movement (orange); (**b**) knee movement.

Figure 2b gives the dynamic measurement of the knee moving from full extension to fully bent for two different knit implementations. The crinkle-related memory effect is apparent at the start. It is slightly larger for the rib than the jersey because the jersey is firmer due to the smaller needle size used. The 1/1 rib implementation gives the largest variation of L for the same change in angle. This difference might be due to needle size rather than stitch type. Smaller needles give a firmer knit and allows less relative change between windings.

4. Electronic Readout

The block diagram of a possible readout system is given in Figure 3a. This is based on a high-Q oscillator as proposed in [8]. The time domain response of the oscillator is simulated using SPICE. The output of the oscillator for the two knee sleeve implementations in their extreme positions is given in Figure 3b.

The high-Q oscillator ensures a readable output for the digital block even with the low-Q knitted coils. The low Q of the coils causes start up delays, but at $\tau < 100$ μs it does not interfere with the operation of the digital readout block where a waiting time of 5 ms is implemented to settle the counter [8]. These delays are directly related to the coils' Q. The performance parameters of coils and oscillator are given in Table 1. The maximum variation of L for knee sleeve 1 is $\Delta L = 104$ μH and 167 μH for sleeve 2. This translates in a frequency variation of 144.7 kHz and 100.4 kHz, respectively. Although τ is largest for sleeve 1, this implementation is preferred over sleeve 2 from a readout point of view.

In conclusion, inductive 3D knitted coils integrated in sleeves of garments give unobtrusive sensors in knitted garments. Their high sensitivity is related to the fast increase in self-inductance with decreasing separation between windings. A maximum inductance variation of 167 μH was measured for the knee moving from full stretch to fully bent. This change can be measured in real time using a high-Q oscillator. The knitted coils are a feasible approach for real-time ambulatory monitoring using fully wearable everyday-looking clothes.

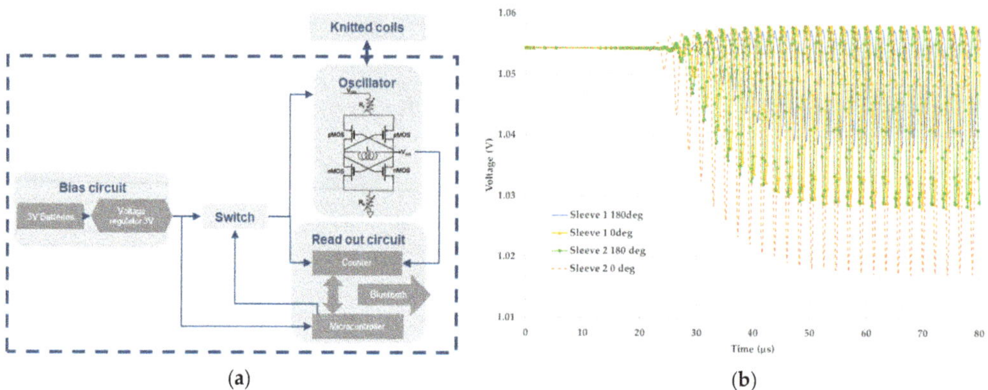

Figure 3. (**a**) Block diagram of an electronic readout. The oscillator has low capacitance FETs to boost its quality factor Q [8]. (**b**) Oscillator response for the knee sleeves in two extreme situations: fully stretched 180° and fully bent ~0°.

Table 1. Performance parameters of the oscillator in the read-out circuit for knee sleeves. L inductance, R series resistance of the coil, f_{osc} the oscillation frequency, $|A|$ the amplitude and τ the start-up delay.

	Angle (°)	L (μH)	R (Ω)	f_{osc} (kHz)	$|A|$ (mV)	τ (μs)
Sleeve 1	180	188	107	744.5	20	70
Sleeve 1	~0	292	107	599.8	30	58
Sleeve 2	180	338	124	558.0	30	50
Sleeve 2	~0	505	124	457.6	41	49

Author Contributions: Conceptualization, K.F.; methodology, K.F.; software, C.P.; validation, K.F. and C.P.; formal analysis, K.F.; investigation, K.F.; resources, K.F.; data curation, K.F. and C.P.; writing—original draft preparation, K.F.; writing—review and editing, K.F. and C.P.; visualization, K.F.; supervision, K.F.; project administration, K.F.; funding acquisition, NA. All authors have read and agreed to the published version of the manuscript.

Funding: No external funding was received for this work.

Institutional Review Board Statement: Ethical review and approval were waived for this study as the dataset is not related to clinical trials, is fully anonymized and consent was given.

Informed Consent Statement: Informed consent was obtained from all subjects involved in the study.

Data Availability Statement: Data can be requested from K.F.

Conflicts of Interest: The authors report no conflict of interest.

References

1. Simpson, L.A.; Menon, C.; Hodgson, A.J.; Ben Mortenson, W.; Eng, J.J. Clinicians' perceptions of a potential wearable device for capturing upper limb activity post-stroke: A qualitative focus group study. *J. Neuroeng. Rehabil.* **2021**, *18*, 135. [CrossRef]
2. Zhou, H.; Stone, T.; Hu, H.; Harris, N. Use of multiple wearable inertial sensors in upper limb motion tracking. *Med. Eng. Phys.* **2008**, *30*, 123–133. [CrossRef]
3. Isaia, C.; McMaster, S.A.; McNally, D. Study of Performance of Knitted Conductive Sleeves as Wearable Textile Strain Sensors for Joint Motion Tracking. In Proceedings of the 2020 42nd Annual International Conference of the IEEE Engineering in Medicine & Biology Society (EMBC), Montreal, QC, Canada, 20–24 July 2020; pp. 4555–4558.
4. Bonroy, B.; Meijer, K.; Dunias, P.; Cuppens, K.; Gransier, R.; Vanrumste, B. Ambulatory monitoring of physical activity based on knee flexion/extension measured by inductive sensor technology. *Int. Sch. Res. Not.* **2013**, *2013*, 908452. [CrossRef]

5. Fobelets, K.; Thielemans, K.; Mathivanan, A.; Papavassiliou, C. Characterization of knitted coils for e-textiles. *IEEE Sens. J.* **2019**, *19*, 7835–7840. [CrossRef]
6. Li, M.; Torah, R.; Nunes-Matos, H.; Wei, Y.; Beeby, S.; Tudor, J.; Yang, K. Integration and Testing of a Three-Axis Accelerometer in a Woven E-Textile Sleeve for Wearable Movement Monitoring. *Sensors* **2020**, *20*, 5033. [CrossRef]
7. Fobelets, K.; Sareen, K.S.; Thielemans, K. Magnetic coupling with 3D knitted helical coils. *Sens. Actuators A Phys.* **2021**, *332*, 113213. [CrossRef]
8. Kiener, K.; Anand, A.; Fobelets, W.; Fobelets, K. Low Power Respiration Monitoring using Wearable 3D knitted Helical Coils. *IEEE Sens. J.* **2022**, *22*, 1374–1381. [CrossRef]

Proceeding Paper

Respiratory Inductive Plethysmography System for Knitted Helical Coils [†]

Kevin Kiener [1,2], Aishwarya Anand [1], William Fobelets [3] and Kristel Fobelets [1,*]

1. Electrical and Electronic Engineering, Imperial College London, Exhibition Road, London SW7 2BT, UK; kevin.kiener@freenet.de (K.K.); aishwarya.anand20@imperial.ac.uk (A.A.)
2. Electrical and Electronic Engineering, Technical University of Munich, 80333 Munich, Germany
3. Barfo, Brusselse Straat 144, 1750 Lennik, Belgium; William.fobelets@scarlet.be
* Correspondence: k.fobelets@imperial.ac.uk
† Presented at the 3rd International Conference on the Challenges, Opportunities, Innovations and Applications in Electronic Textiles (E-Textiles 2021), Manchester, UK, 3–4 November 2021.

Abstract: Three-dimensional knitted helical coils are very sensitive inductive sensors that can be used to monitor breathing. Their inductance is high and the quality factor relatively low. A read-out circuit is designed and tested to track the inductance variations during circumference changes of a phantom chest. The challenge of the low-quality factors of the coil is resolved by designing a double cross-coupled FET pair with low capacitance. A digital counter records the frequency. A microprocessor samples the signal every 250 ms to minimize power consumption.

Keywords: high-Q oscillator; inductive sensor; breathing monitoring

1. Introduction

Inductive sensors can be used in a variety of environments, one of which is non-contact body sensing networks. Application areas include respiration monitoring [1], ambulatory monitoring [2] and wireless power transfer [3]. Sensors require not only appropriate sensing characteristics but also readout electronics to transform the signal into practical data for further processing. In addition, the electronics need to run on batteries for long periods of time. In this work, we propose a low-power readout circuit for a 3D knitted helical coil. The coils have a high sensitivity up to ~68 nH/mm circumference change, a high rest inductance ~40 µH and a high resistance ~20 Ω, depending on knit implementation. The high resistance makes the quality factor Q of the coil relatively low. We present a high-Q low-power oscillator solution that gives a fast and stable response. A frequency counter converts the analogue signal into digital, and a microcontroller controls the sampling rate. This implementation reads breathing signals with good accuracy and at low power levels.

2. Materials and Methods

Figure 1 illustrates the knitted inductive sensor. This is created by knitting thin insulated metal wire and ordinary yarn simultaneously. A circular knit must be implemented to form a helical coil [4]. Figure 1a is a picture of the knit stitches with grey 3 mm yarn and red Cu insulated wire (~250 µm diameter). Figure 1b is a sample coil on a small chest phantom with moving "ribs". For dynamic measurements, the yellow disk is rotated. The speed and angle of the rotation is controlled by a computer. Plastic slides are inserted next to the oval disk for static measurements.

Citation: Kiener, K.; Anand, A.; Fobelets, W.; Fobelets, K. Respiratory Inductive Plethysmography System for Knitted Helical Coils. *Eng. Proc.* 2022, *15*, 7. https://doi.org/10.3390/engproc2022015007

Academic Editors: Steve Beeby, Kai Yang and Russel Torah

Published: 11 March 2022

Publisher's Note: MDPI stays neutral with regard to jurisdictional claims in published maps and institutional affiliations.

Figure 1. (**a**) Thin insulated metal wire integrated in the knit stitches that form the 3D helical coil. (**b**) The chest phantom holding a test coil. The oval disk is connected to a small motor to rotate it around its axis over programmed angles and with a given speed.

The circuit diagram of the readout is given in Figure 2. All components are defined on the diagram. The oscillator is formed by two cross-coupled FET pairs [5]. The top pair use pMOS to supply power to the bottom pair. The bottom nMOS pair form a negative resistance. The variable inductor represents the coil. No further components are added as the capacitance of the FETs forms the LC oscillator with the coil's inductance. The capacitance of the cross-coupled pair is ~41pF, depending on the chosen bias point of the FETs. FETs should remain in their linear region to limit harmonics. Low C FETs boost the quality factor of the oscillator $Q = \frac{1}{R}\sqrt{\frac{L}{C}}$, with R as the internal coil resistance, L as its inductance and C as the FETs' capacitance. The power supply to the oscillator is stabilized using a low-power 5 V voltage regulator. This ensures that the variation of the battery voltage upon use is minimized and does not interfere with the oscillation frequency.

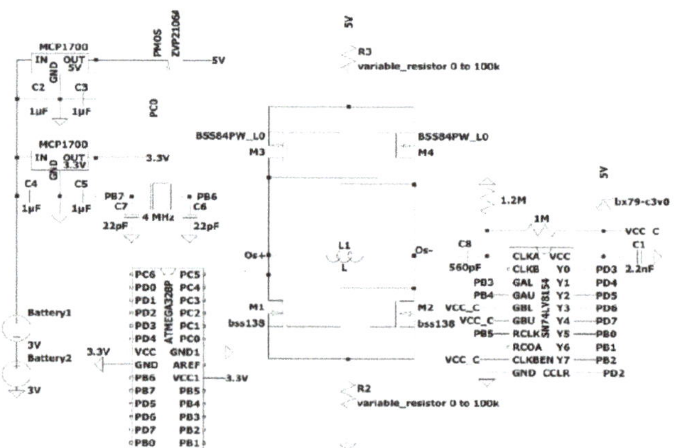

Figure 2. Circuit diagram. All components are identified.

In this implementation, two variable resistors were used. The top one controls current flow and thus amplitude of the oscillations. The bottom resistor controls the DC offset from the ground. The input signal for a logical one of the counters must be higher than 0.7 × Vcc and lower than 0.3 × Vcc for a logical zero. Thus, the resistor values are chosen so as to ensure correct operation of the counter and to minimize its current consumption. These settings are linked to the specific coil used and can be fixed once the coil type is fixed. Since the counter and microcontroller are power-hungry, their supply is minimized to ~3 V. A pMOS switch (Q1 in Figure 2), controlled by the microcontroller, switches the circuit on and off. The system is on for ~45 ms and in sleep mode for ~250 ms. The frequency is counted for 10 ms and output recorded by the microcontroller. The oscillator response time is less than 10 µs; however, the counter takes longer and a ~5 ms delay in recording the output is implemented. To further reduce power, the microprocessor runs on an external clock of 4 MHz. This implementation allows the circuit to run on the same batteries for multiple days.

3. Results

For testing, the circuit was implemented on a breadboard (Figure 3), with the oscillator and counter implemented on a PCB. Realizing the full circuit on a PCB with surface mount components and a microcontroller with integrated counter and watchdog timer will make the implementation more compact and wearable.

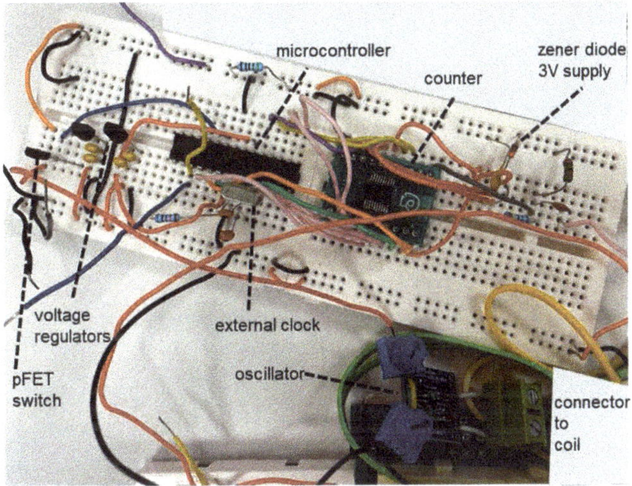

Figure 3. The circuit of Figure 3 implemented on a breadboard.

The performance of the circuit was tested using the dynamic setting of the phantom chest and rotating the disk over ~60° at a speed of ~2.4 turns/s. To analyse whether the circuit gives a true representation of the "breathing" signal, the measured oscillation frequency was converted into an inductance using $C = 40.8$ pF. This measurement can then be easily compared to the measurement of the same breathing setup with a Wayne–Kerr 6500B precision component analyser. Figure 4a shows the inductance changes using the circuit and Figure 4b that of the precision component analyser. The measured variations are similar, but the details of the geometrical profile changes in the disk (see the higher inductance values) are less well-represented. This is because of the lower sampling rate of the circuit compared to the Wayne–Kerr. Increasing the sampling rate is possible but will consume more power and reduce battery life. Overall, our low-power implementation gives a fair representation of the simulated breathing signal.

Different coils were knitted in a 1/1 rib stitch using cotton yarn with a different number of rows that include the metal wire. The number of knitted rows with the wire is equivalent to the number of windings (turns) in a coil. All coils had the same diameter. Increasing the number of turns, N increases the self-inductance of the coil but also increases its resistance. The oscillation frequency $f = \frac{1}{2\pi\sqrt{LC}}$ decreases with the increasing number of turns. This translates into a decrease in power consumption with the increasing N. Figure 5a gives the power consumed by the oscillator and counter during the time it is on but ignores the power consumed by the microprocessor that remains the same for all implementations. Figure 5b gives the sensitivity: $\Delta f / \Delta l$ with l circumference. This demonstrates a trade-off between power and sensitivity. This trade-off is a consequence of converting inductance to frequency that causes a decrease in sensitivity for an increase in rest inductance value.

For the given coil implementation, a compromise would be to use $N = 5$, that has a higher power consumption but also higher sensitivity.

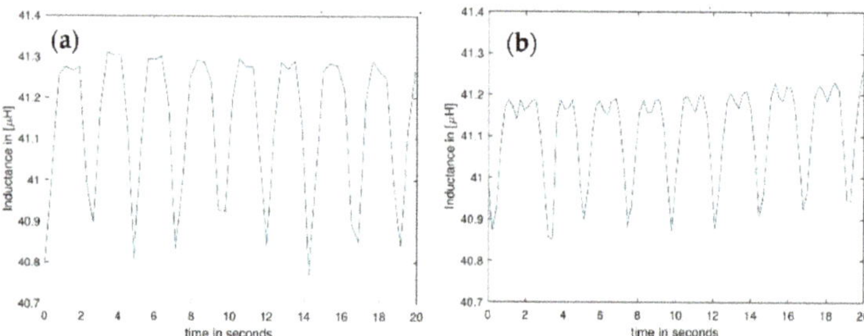

Figure 4. Result of a dynamic measurement using the phantom chest (**a**) using the readout circuit given in Figure 3 and converting the frequency back to inductance changes and (**b**) using the precision impedance amplifier (Wayne–Kerr).

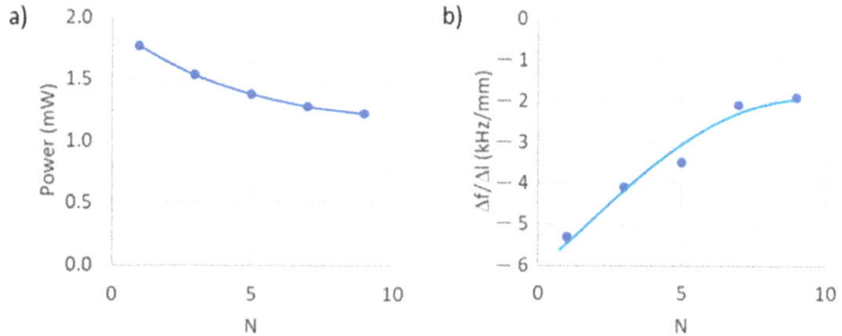

Figure 5. (**a**) Power consumed by the circuit (excluding the microprocessor) as a function of the number of turns N. (**b**) The variation in the frequency as a function of variation of circumference l. The knit is a 1/1 rib knit in cotton. The chest phantom of Figure 2 is used for static measurements using the thin plastic sliders to increase circumference.

To decrease power consumption, the total length of the metal wire integrated in the knit can be reduced by reducing the number of stitches in which the metal is integrated. This needs to be done carefully so as not to jeopardize the flexibility in the knit. To increase sensitivity, a jersey rather than a rib stitch can be used, as a jersey stitch gives a lower inductance value for the same number of turns. In addition, it was found that the yarn crossover points between the stitches in jersey knit increase the inductance variation as a function of stretch compared to that using rib stitch. A more tight-fitting garment obtained by reducing needle size will also boost performance as it not only decreases the metal wire length but also follows chest movements better [6].

In conclusion, the design of a high-Q oscillator using low-capacitance FETs in a double cross-coupled pair configuration allows the readout of breathing signals sensed with 3D knitted helical coils that have a high sensitivity in inductance variation with circumference but also a high resistance. This supports the implementation of fully wearable knitted breathing sensors.

Author Contributions: Conceptualization, K.F. and W.F.; methodology, K.F. and K.K.; validation, K.K., A.A. and K.F.; formal analysis, K.K., A.A. and K.F.; investigation, K.K. and A.A.; resources, K.F.; data curation, K.F.; writing—original draft preparation, K.F.; writing—review and editing, K.K., A.A. and K.F.; visualization, K.F.; supervision, K.F.; project administration, K.F. All authors have read and agreed to the published version of the manuscript.

Funding: This research received no external funding.

Institutional Review Board Statement: Not applicable.

Informed Consent Statement: Not applicable.

Data Availability Statement: From K.F. upon request.

Acknowledgments: We thank V. Body and A. Halimi for help in the lab.

Conflicts of Interest: The authors declare no conflict of interest.

References

1. Fobelets, K. Knitted coils as breathing sensors. *Sens. Actuators A Phys.* **2020**, *306*, 111945. [CrossRef]
2. Fobelets, K. Ambulatory Monitoring Using Knitted 3D Helical Coils. In Proceedings of the 3rd International Conference on the Challenges, Opportunities, Innovations and Applications in Electronic Textiles (E-Textiles 2021), Manchester, UK, 3–4 November 2021.
3. Fobelets, K.; Sareen, K. Thielemans K, Magnetic coupling with 3D knitted helical coils. *Sens. Actuators A Phys.* **2021**, *332*, 113213. [CrossRef]
4. Fobelets, K.; Thielemans, K.; Mathivanan, A.; Papavassiliou, C. Characterisation of knitted coils for e-textiles. *IEEE Sens. J.* **2019**, *19*, 7835–7840. [CrossRef]
5. Ebrahimzadeh, M. Design of an ultra-low power low phase noise CMOS LC oscillator. *Int. J. Soft Comput. Eng. (IJSCE)* **2011**, *1*, 78–81.
6. Kiener, K.; Anand, A.; Fobelets, W.; Fobelets, K. Low Power Respiration Monitoring using Wearable 3D knitted Helical Coils. *IEEE Sens. J.* **2022**, *22*, 1374–1381. [CrossRef]

engineering proceedings

MDPI

Proceeding Paper

Textile Tactile Senor Based on Ferroelectret for Gesture Recognition †

Junjie Shi * and Mahmoud Wagih

Smart Electronic Materials and Systems Group, School of Electronics and Computer Science,
University of Southampton, Southampton SO17 1BJ, UK; mahmoud.wagih.mohamed@soton.ac.uk
* Correspondence: Junjie.Shi@soton.ac.uk

† Presented at the 3rd International Conference on the Challenges, Opportunities, Innovations and Applications
in Electronic Textiles (E-Textiles 2021), Manchester, UK, 3–4 November 2021.

Abstract: Ferroelectret is a charged polymer with cellular void structures that create giant dipole moments across the material's thickness. In this work, we present the first realization of a wearable textile substrate tactile sensor based on Polypropylene (PP) ferroelectret material for gesture recognition. As a result, the sensitivity of the fabricated sensor is 0.21 V/kPa in the pressure range of 0–20 kPa. The ferroelectret tactile sensor adheres to a glove's surface for detecting human movements such as bending or the relaxation motion of the palm and the bending or stretching motion of each finger, enabling the successful detection of small finger gestures around a 400 mV output.

Keywords: ferroelectret; tactile sensor; gesture recognition

1. Introduction

With the explosive growth of digital devices, the technologies enabling interaction between people and devices are required more than ever. Among them, gesture recognition technology has received a lot of research attention [1–3]. In essence, gesture recognitions are divided into two categories: contact sensors and non-contact sensors. Non-contact sensors are mostly based on visual technologies [4,5]. Non-contact sensors utilize various technologies to extract information about the shape and movement of the hand from instantaneous image, and then recognizes gestures based on this information. Alternatively, the most commonly used methods of recognizing gestures by contact sensors are Electromyography (EMG) sensors and force-sensitive resistors sensors (FSRs) [6–8]. However, the drawbacks of these two types of sensors are also obvious. For EMG sensors, the electric potential of muscles is too sensitive relative to electric noise, because the magnitude of the electric potential is in range of submillivolts. FSRs are more robust to noise compared to EMG sensors, but they require a continuous external power supply.

Sensor based on piezoelectric materials such as Zinc Oxide (ZnO), lead zirconate titanate (PZT), Molybdenum disulfide (MoS$_2$), poly(vinylidene fluoride) (PVDF) and its similar copolymer p(vdf-TrFE) exhibit piezoelectricity by converting mechanical signals into electrical outputs. For these conventional piezoelectric materials, PZT and PVDF, PZT possesses excellent dielectric and piezoelectric properties, owing to its very high Young's modulus (63 GPa). However, PZT is a rigid ceramic unsuitable for wearable e-textile applications [9]. In contrast, PVDF is a soft polymer with a low Young's modulus (2.9 GPa), and its piezoelectric charge coefficient d_{33} is an order of magnitude lower than of PZT (20pC/N), which is insufficient for certain applications that require high bending and high sensitivity requirements [9].

A ferroelectret is a thin film of polymer foam that can generate an electrical signal under mechanical force, similarly to piezoelectric materials. However, the piezoelectric effect in a ferroelectret is very different from piezoelectric materials. Its piezoelectricity comes from the separated positive and negative charges, which are trapped on the upper

Citation: Shi, J.; Wagih, M. Textile Tactile Senor Based on Ferroelectret for Gesture Recognition. *Eng. Proc.* **2022**, *15*, 8. https://doi.org/10.3390/engproc2022015008

Academic Editors: Steve Beeby, Kai Yang and Russel Torah

Published: 11 March 2022

Publisher's Note: MDPI stays neutral with regard to jurisdictional claims in published maps and institutional affiliations.

and lower surface of the voids [10]. Upon applying pressure or bending, the change in the dipole moments generates a change in the accumulated electric charge on each surface of ferroelectret. Due to its outstanding properties in comparison with other piezoelectric material, numerous applications have been proposed, such as acoustic transducers, higher frequency loudspeakers and keyboards [10,11].

In this paper, a wearable ferroelectret tactile sensor based on PP for gesture recognition is presented. The proposed tactile sensor was fabricated based on PP ferroelectret with a polydimethylsiloxane (PDMS) encapsulation on a textile substrate and shows a high sensitivity to pressure, which can be utilized to monitor the movement of hands.

2. Sensor Fabrication

The schematic diagram of the fabrication processes of a ferroelectret tactile sensor is illustrated in Figure 1a. PP film sheets were commercially purchased from EMFIT. These films are cut into sample sizes that can be attached to a glove. A thin layer of silver electrodes was screen-printed on the top and bottom of the PP ferroelectret, respectively. These samples with printed silver electrodes need to be cured in an oven at 50 °C for 10 min. In order to obtain the sensor signal, a small piece of conductive tape was used to connect external wires to these electrodes. After that, a thin layer of PDMS was casted on the electrodes by spin coating as an encapsulation layer. The samples with the PDMS encapsulation layer were heated in an oven at 50 °C for 10 min, and the sample was then left at room temperature for one day to fully cure. The fabricated ferroelectret tactile sensor was bonded to a textile substrate (glove) by a thin layer of adhesive tape, as shown in Figure 1b.

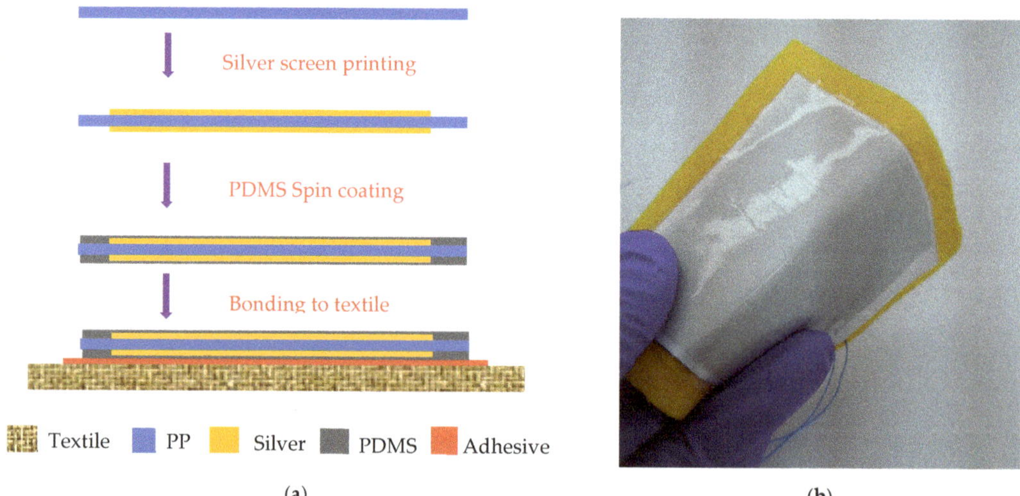

(a)   (b)

Figure 1. (**a**) The fabrication process for the ferroelectret tactile sensor; (**b**) the photo of the fabricated tactile sensor.

3. Experiments and Results

To demonstrate the feasibility of using this sensor for gesture recognition, the fabricated tactile sensors were sequentially installed in each finger and palm to monitor the movement of these areas, as shown in Figure 2. An oscilloscope was used to record the voltage generated from this ferroelectret tactile sensor.

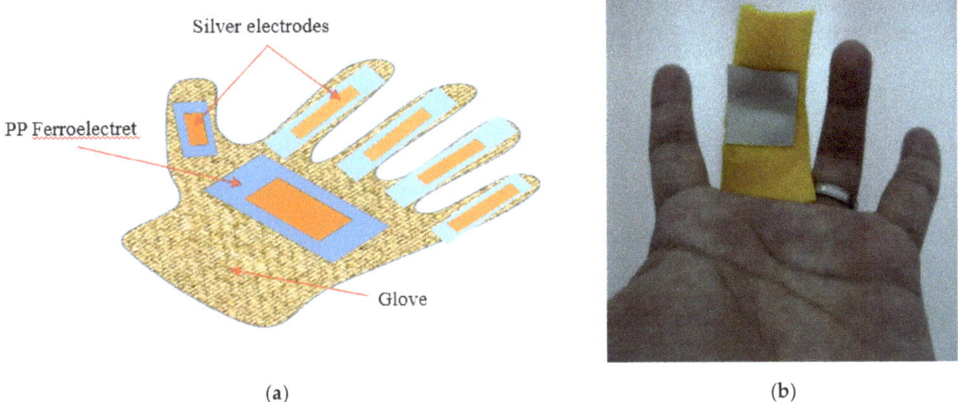

(**a**)

(**b**)

Figure 2. (**a**) The schematic of diagram of the ferroelectret tactile sensor system for gesture recognition; (**b**) the image of ferroelectret tactile sensor for finger bending.

During a fist gesture, the output voltages of the tactile sensor installed at different hand positions are in the form of a voltage pulse, as shown in Figure 3. The maximum output voltages of the sensors from the middle finger, pinky finger and palm positions are 0.8 V, 0.4 V and 0.9 V, respectively. In particular, the durations of the output voltage spike for different tactile sensor positions are significantly different.

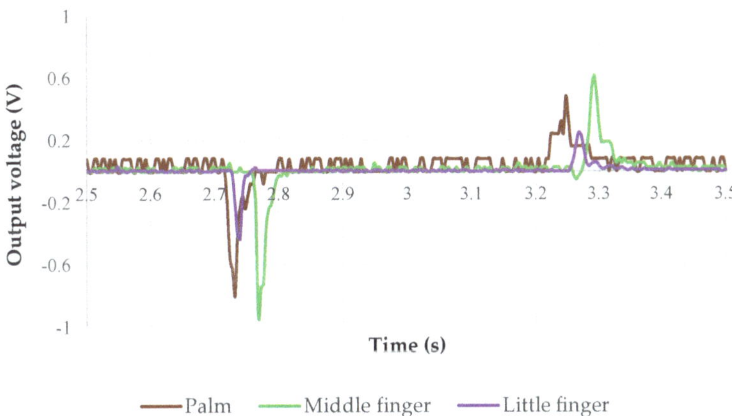

Figure 3. The output voltage of the tactile sensor installed at fingers and palm during a fist gesture.

To understand the sensing mechanism for gesture recognition, the output voltage curve of the ferroelectret sensor during the entire process from bending to the relaxation of the finger is shown in Figure 4a. Compared with the process of releasing, maximum voltage amplitude increases to 0.93 V during finger bending. Afterwards, the voltage returns to its original value when no further pressure is applied. A maximum voltage amplitude 0.61 V was measured when releasing. To quantify the sensitivity of the fabricated tactile sensor, the sensor was tested under an external pressure applied by an electrodynamic instrument (ElectroPuls E1000, Instron Ltd., Buckinghamshire, UK). The "open circuit" peak voltage, measured across a 10 MΩ probe impedance, as a function of external pressure is shown in Figure 4b. There is a direct linear relationship between the measured open voltage and the external pressure. The sensitivity of the tactile sensor is 0.21 V/kPa in the pressure range of 0–20 kPa.

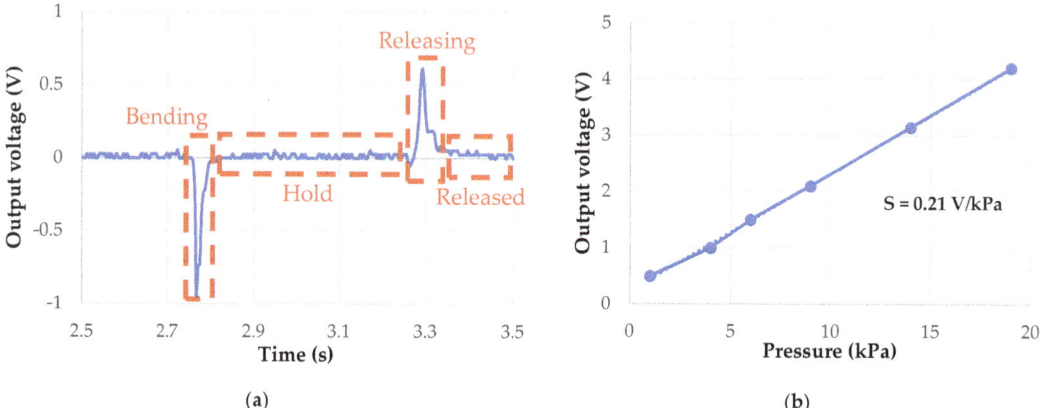

Figure 4. (**a**) Output voltage of the ferroelectret tactile sensor under different states for a finger bending; (**b**) output voltage response under different pressure.

4. Conclusions

In this work, we presented the first realization of a wearable tactile sensor implemented on a textile substrate based on PP ferroelectret material for gesture recognition. The sensitivity of the fabricated sensor was achieved at 0.21 V/kPa in a pressure range of 0–20 kPa. The ferroelectret tactile sensor demonstrates the feasibility of detecting human movement, such as bending/relaxation motions of the palm and bending/stretching motion of each finger, and can successfully detect pinky finger gestures with an output of about 400 mV.

Author Contributions: Conceptualization, J.S. and M.W.; designed, implemented the device and conducted the experiment J.S.; J.S. and M.W. wrote the paper. All authors have read and agreed to the published version of the manuscript.

Funding: This research received no external funding.

Institutional Review Board Statement: Not applicable.

Informed Consent Statement: Not applicable.

Data Availability Statement: Not applicable.

Acknowledgments: This work was performed under the "Wearable and Autonomous Computing for Future Smart Cites", a Platform Grant funded by the UK engineering and Physical Sciences Research Council (EPSRC), Grant EP/P010164/1.

Conflicts of Interest: The authors declare no conflict of interest.

References

1. Xu, C.; Pathak, P.H.; Mohapatra, P. Finger-writing with Smartwatch: A Case for Finger and Hand Gesture Recognition using Smartwatch. In Proceedings of the 16th International Workshop on Mobile Computing Systems and Applications (Hotmobile' 15), Santa Fe, NM, USA, 12–13 February 2015; pp. 9–14.
2. McIntosh, J.; McNeill, C.; Fraser, M.; Kerber, F.; Lochtefeld, M.; Kruger, A. EMPress: Practical Hand Gesture Classification with Wrist-Mounted EMG and Pressure Sensing. In Proceedings of the 34th Annual Chi Conference on Human Factors in Computing Systems, Chi 2016, San Jose, CA, USA, 7–12 May 2016; pp. 2332–2342.
3. Zhang, Y.; Harrison, C. Tomo: Wearable, Low-Cost, Electrical Impedance Tomography for Hand Gesture Recognition. In Proceedings of the Uist'15: Proceedings of the 28th Annual Acm Symposium on User Interface Software and Technology, Charlotte, NC, USA, 11–15 November 2015; pp. 167–173.
4. Lv, Z.H.; Halawani, A.; Feng, S.Z.; ur Rehman, S.; Li, H.B. Touch-less interactive augmented reality game on vision-based wearable device. *Pers. Ubiquitous Comput.* **2015**, *19*, 551–567. [CrossRef]

5. Zhao, R.N.; Ma, X.L.; Liu, X.H.; Li, F.M. Continuous Human Motion Recognition Using Micro-Doppler Signatures in the Scenario with Micro Motion Interference. *IEEE Sens. J.* **2021**, *21*, 5022–5034. [CrossRef]

6. Cheng, Y.; Li, G.; Yu, M.; Jiang, D.; Yun, J.; Liu, Y.; Liu, Y.; Chen, D. Gesture recognition based on surface electromyography-feature image. *Concurr. Comput. Pract. Exp.* **2021**, *33*, e6051. [CrossRef]

7. Wu, Y.H.; Chen, K.; Fu, C.L. Natural Gesture Modeling and Recognition Approach Based on Joint Movements and Arm Orientations. *IEEE Sens. J.* **2016**, *16*, 7753–7761. [CrossRef]

8. Esposito, D.; Andreozzi, E.; Gargiulo, G.D.; Fratini, A.; D'Addio, G.; Naik, G.R.; Bifulco, P. A Piezoresistive Array Armband with Reduced Number of Sensors for Hand Gesture Recognition. *Front. Neurorobotics* **2020**, *13*, 114. [CrossRef] [PubMed]

9. Komolafe, A.; Zaghari, B.; Torah, R.; Weddell, A.S.; Khanbareh, H.; Tsikriteas, Z.M.; Vousden, M.; Wagih, M.; Jurado, U.T.; Shi, J.J.; et al. E-Textile Technology Review-From Materials to Application. *IEEE Access* **2021**, *9*, 97152–97179. [CrossRef]

10. Zhang, Y.; Bowen, C.R.; Ghosh, S.K.; Mandal, D.; Khanbareh, H.; Arafa, M.; Wan, C.Y. Ferroelectret materials and devices for energy harvesting applications. *Nano Energy* **2019**, *57*, 118–140. [CrossRef]

11. Rupitsch, S.J.; Lerch, R.; Strobel, J.; Streicher, A. Ultrasound Transducers Based on Ferroelectret Materials. *IEEE Trans. Dielectr. Electr. Insul.* **2011**, *18*, 69–80. [CrossRef]

engineering proceedings

 check for updates

Citation: Wagih, M.; Malik, O.; Weddell, A.S.; Beeby, S. E-Textile Breathing Sensor Using Fully Textile Wearable Antennas. *Eng. Proc.* **2022**, *15*, 9. https://doi.org/10.3390/engproc2022015009

Academic Editors: Russel Torah and Kai Yang

Published: 15 March 2022

Publisher's Note: MDPI stays neutral with regard to jurisdictional claims in published maps and institutional affiliations.

Proceeding Paper

E-Textile Breathing Sensor Using Fully Textile Wearable Antennas [†]

Mahmoud Wagih *[iD], Obaid Malik, Alex S. Weddell [iD] and Steve Beeby [iD]

School of Electronics and Computer Science, University of Southampton, Southampton SO17 1BJ, UK; om1e09@ecs.soton.ac.uk (O.M.); asw@ecs.soton.ac.uk (A.S.W.); spb@ecs.soton.ac.uk (S.B.)
* Correspondence: mahm1g15@ecs.soton.ac.uk
† Presented at the 3rd International Conference on the Challenges, Opportunities, Innovations and Applications in Electronic Textiles (E-Textiles 2021), Manchester, UK, 3–4 November 2021.

Abstract: E-textile sensor networks enable a variety of applications including pervasive monitoring for distributed healthcare. While commercial wearables can now measure various quantities such as heart rate and activities in a real-time, robust, and pervasive manner, breathing sensors remain an ongoing research challenge. In this paper, the use of wearable antennas for respiration monitoring is investigated based on a low-profile broadband fully textile antenna. It is demonstrated that the antenna, suitable for operation on different substrates and body parts, exhibits over 2 dB wireless gain sensitivity to normal breathing. Unlike recent wearable breathing sensors, the proposed antenna has a very simple structure and does not rely on active mechanical sensing elements or specific materials. A simple peak-detection algorithm is investigated showing a nearly 100% breath detection accuracy in line-of-sight. Based on the experimental results, it can be concluded that e-textile antennas can be utilized as highly accurate sensors for respiration monitoring, without the need for specific sensing elements or materials.

Keywords: wearable sensor; antenna sensor; RF breathing sensor; wireless sensors; textile sensors; vital sign monitoring

1. Introduction

Owing to our ageing population, remote healthcare technologies have attracted significant research interest. At the forefront of enabling technologies is wireless vital sign monitoring, where body-centric wearable sensors could report information such as the heart rate or breathing rate of a subject, yielding valuable information for evaluating their health condition [1].

Electronic textiles (e-textiles) are the closest non-invasive interface to the user, where textile-based sensors can be used to monitor a variety of parameters including daily activities, vital signs [1], as well as environmental conditions such as moisture [2]. Furthermore, extensive research efforts have demonstrated methods of reliably connecting and powering e-textile systems using wearable antennas for information and power transmission [3], as well as using on-body energy storage devices, capable of powering low-power sensor nodes based on off-the-shelf components such as Bluetooth transceivers [4].

With the increased popularity of radio frequency identification (RFID) as a ubiquitous technology in retail and asset-tracking, several research efforts have demonstrated passive RFID sensors [5]. In a passive RFID or antenna-based sensor, the parameter-under-test (PUT) is detected through a change in the antenna's gain, which can be determined through the received signal strength indicator (RSSI) value [5]. RFID sensors have several advantages, such as wireless battery-free operation, where they harvest RF power from the reader, as well as being ultra-low-cost systems, with the only active electronic component being the RFID integrated circuit (IC).

In this paper, we propose a flexible all-textile broadband antenna as a passive e-textile breathing sensor. Unlike existing approaches where the antenna relies on a special fabric or material to realize a stretchable or compressible element, the proposed sensing approach relies on the human body-antenna interaction. The sensor is experimentally characterized and combined with a simple peak-detection algorithm for post-processing, demonstrating high accuracy in detecting breaths.

2. E-Textile Sensing Antenna

To realize an e-textile sensing antenna for vital sign monitoring, the antenna needs to maintain its operation in proximity with the human body. This is evaluated through the antenna's radiation efficiency and reflection coefficient, where a sufficiently low reflection (under 10%) indicates that the antenna is suitable for integration with different transmitters. Moreover, the antenna needs to be unisolated from the human body in order to detect the changes in dielectric properties, induced by breathing, through its gain. Therefore, the sensing antenna is based on a broadband monopole antenna, previously demonstrated with relatively high gain and efficiency on the body [6]. Furthermore, as the antenna is low-profile and does not require a specific substrate thickness to radiate efficiently, the selected design is suitable for implementation on different textile substrates. Figure 1a,b show the layout and photograph of the antenna, respectively.

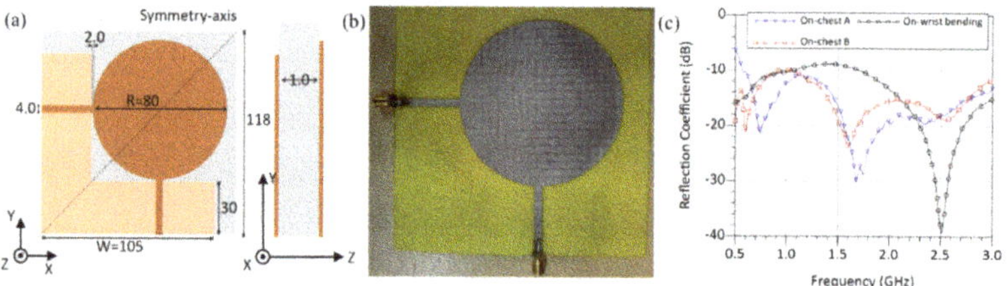

Figure 1. The proposed e-textile sensing antenna: (**a**) layout and dimensions (in mm); (**b**) photograph of the fabricated prototype; (**c**) measured reflection coefficient of the antenna on different body parts showing a matched impedance.

The antenna was fabricated using conductive fabric from P&P Technology (Metweave), laser cut to the antenna's dimensions. The conductive fabric was then attached to the textile substrate using its adhesive backing. The substrate used in this work was felt fabric, which maintains a relative permittivity of 1.2 and a dielectric loss dissipation factor of 0.02. The antenna's impedance bandwidth was experimentally characterized to demonstrate its suitability for interrogation across different frequency bands. The reflection coefficient was measured using a Rohde and Schwarz ZVB4 vector network analyzer (VNA) calibrated using a standard Through, Open, Short, Match (TOSM) calibration. Sub-miniature type-A (SMA) connectors were soldered directly onto the conductive fabric microstrip feed, as shown in the photograph in Figure 1b. Figure 1c shows the measured reflection coefficient of the antenna from 0.5 to 3 GHz, for different on-body positions, where the antenna was placed on both sides of the user's chest. For the breathing sensor application, the antenna will be mounted on the user's chest, where it can be observed that under −10 dB reflections were measured. Therefore, the antenna can be interrogated wirelessly in either the sub-1 GHz (868/915 MHz) or 2.4 GHz license-free bands, for wider compatibility with off-the-shelf RFID readers as well as Bluetooth or Wi-Fi transceivers.

3. Breathing Sensor Characterization

The proposed antenna-based sensor relies on the gain to detect the PUT, i.e., the breathing rate. To explain, as the user inhales, the antenna's gain increases, which is attributed to the reduction in the dielectric losses in the antenna. The antenna's gain, in-turn, affects the measured channel gain which can be detected wirelessly using the RSSI value. To evaluate the antenna's response as a sensor, the VNA was used to measure the channel gain through the forward transmission (S21) between the textile sensing antenna on the user's chest and a reference directional antenna as a transmitter. The transmitting and receiving antennas were separated by distance d = 1.0 m. The experimental setup is illustrated in Figure 2a.

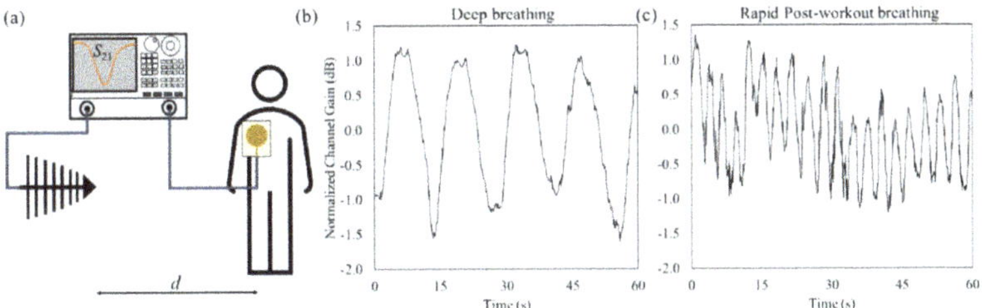

Figure 2. Antenna gain measurements showing the breathing patterns: (**a**) measurement setup; (**b**) measured channel gain during deep breathing; (**c**) measured channel gain during rapid breathing.

The channel gain was measured over a period of 60 s. Two breathing patterns were measured: slow, deep breathing, and rapid breathing post-activity. Figure 2b,c show the measured channel gain response for both breathing patterns. Each peak observed corresponds to a single breath, where the total number of counted breaths, from the RF sensor, matches the manual breath count. To further demonstrate the suitability of the proposed sensing method for detecting the breathing rate of the user, the measured traces were post-processed using a simple peak detection algorithm. Figure 3 shows the successful breath-rate detection using the peak detection algorithm.

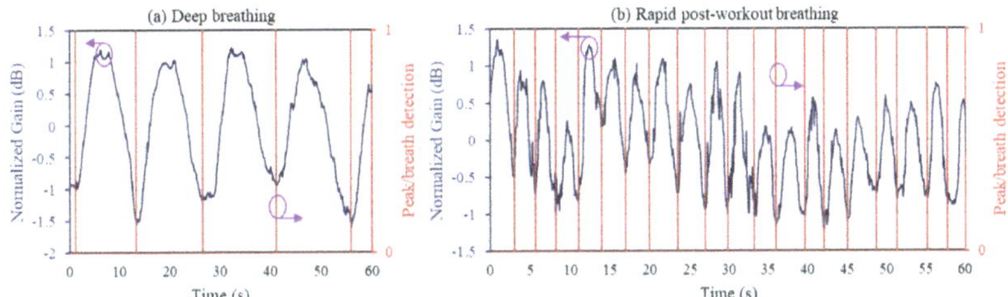

Figure 3. Successful breath rate measurement using a peak-detection algorithm.

The proposed sensor is compared in Table 1 to recently reported wearable breathing sensors. While it can be seen that other sensors have been demonstrated based on e-textile antennas with a higher gain change in-response to breathing [7,8], the antennas in previous works required stretchable or compressible elements to detect the breathing rate through a sufficiently large gain change (in dB). The variation in the measurement range in the compared studies can be linked to the sensitivity of the wireless receiver, where the maximum read range of the sensor is determined by the sensitivity of the reader, as well as the transmitted power level, limited by the radiated power regulations.

Table 1. Comparison of the proposed e-textile breathing sensor with related work.

	This Work	**[1]**	**[7]**	**[8]**
Antenna Material	Textile (felt)	PCB	Stretchable textile	Textile + compressible foam
Sensing mechanism	Dielectric sensing	Dielectric sensing	Strain sensing	Compression sensing
Sensitivity	~1.5–2.5 dB	~10%	1.5–7 dB	3–9 dB
Data post-processing	Peak-detection	N/A	N/A	N/A
Specific fabric required	No	Non-textile	Yes	Yes
Measurement range (m)	1.5	1–1.5	0.5	4–4.5

4. Conclusions

In this paper, the use of a broadband wearable textile antenna as a breathing sensor was proposed. The antenna maintains a stable bandwidth over the human body and on-chest, making it suitable for wearable sensing. Combined with a simple peak detection algorithm, the proposed sensor exhibits high accuracy in detecting the breathing rate and pattern of the user. It is concluded that passive wireless breathing sensors can be developed and integrated on different textiles without the need for specific materials or fabrication methods.

Author Contributions: Conceptualization, M.W.; methodology, M.W.; software, O.M.; writing—original draft preparation, M.W.; writing—review and editing, M.W., A.S.W. and S.B.; funding acquisition, S.B. All authors have read and agreed to the published version of the manuscript.

Funding: This work was supported by the UK Engineering and Physical Sciences Research Council (EPSRC) under Grant EP/P010164/1. The work of Steve Beeby was supported by the Royal Academy of Engineering under the Chairs in Emerging Technologies Scheme.

Institutional Review Board Statement: Ethical approval has been obtained from the Faculty of Engineering and Physical Sciences (FEPS) Ethics Committee, University of Southampton, SO17 1BJ, application number: 64814, approved: 26 August 2021.

Informed Consent Statement: Informed consent was obtained from all subjects involved in the study.

Data Availability Statement: The datasets underpinning this work are available from the corresponding author upon request.

Conflicts of Interest: The authors declare no conflict of interest.

References

1. Hui, X.; Kan, E.C. Monitoring vital signs over multiplexed radio by near-field coherent sensing. *Nat Electron.* **2018**, *1*, 74–78. [CrossRef]
2. Chen, X.; He, H.; Khan, Z.; Sydänheimo, L.; Ukkonen, L.; Virkki, J. Textile-Based Batteryless Moisture Sensor. *IEEE Antennas Wirel. Propag. Lett.* **2020**, *19*, 198–202. [CrossRef]
3. Wagih, M.; Hilton, G.S.; Weddell, A.S.; Beeby, S. Dual-Band Dual-Mode Textile Antenna/Rectenna for Simultaneous Wireless Information and Power Transfer (SWIPT). *IEEE Trans. Antennas Propag.* **2021**, *69*, 6322–6332. [CrossRef]
4. Wagih, M.; Hillier, N.; Weddell, A.; Beeby, S. E-Textile RF energy harvesting and storage using organic-electrolyte carbon-based supercapacitors. In Proceedings of the 2021 IEEE MTT-S International Microwave Workshop Series on Advanced Materials and Processes for RF and THz Applications, Chongqing, China, 15–17 November 2021; p. 4.
5. Zhang, J.; Tian, G.Y.; Marindra, A.M.J.; Sunny, A.I.; Zhao, A.B. A Review of Passive RFID Tag Antenna-Based Sensors and Systems for Structural Health Monitoring Applications. *Sensors* **2017**, *17*, 265. [CrossRef] [PubMed]
6. Wagih, M.; Weddell, A.; Beeby, S. Omnidirectional Dual-Polarized Low-Profile Textile Rectenna With Over 50% Efficiency for Sub-μW/cm^2 Wearable Power Harvesting. *IEEE Trans. Antennas Propag.* **2021**, *69*, 2522–2536. [CrossRef]
7. Liu, Y.; Yu, M.; Xia, B.; Wang, S.; Wang, M.; Chen, M.; Dai, S.; Wang, T.; Ye, T.T. E-Textile Battery-less Displacement and Strain Sensor for Human Activities Tracking. *IEEE Internet Things J.* **2021**, *8*, 16486–16497. [CrossRef]
8. Tajin, M.A.S.; Amanatides, C.E.; Dion, G.; Dandekar, K.R. Passive UHF RFID-based Knitted Wearable Compression Sensor. *IEEE Internet Things J.* **2021**, *8*, 13763–13773. [CrossRef] [PubMed]

engineering proceedings

MDPI

Proceeding Paper

Assessing the Validity of a Kinematic Knee Sleeve in a Resistance-Trained Population [†]

Nathan Toon [1,*], Simon McMaster [2], Tom Outram [1] and Mark Faghy [1]

1   Human Sciences Research Centre, College of Science and Engineering, University of Derby, Derby DE22 1GB, UK; t.outram@derby.ac.uk (T.O.); m.faghy@derby.ac.uk (M.F.)
2   Footfalls and Heartbeats (UK) Limited, Nottingham NG7 1FW, UK; simon@footfallsandheartbeats.com
*   Correspondence: n.toon@derby.ac.uk
†   Presented at the 3rd International Conference on the Challenges, Opportunities, Innovations and Applications in Electronic Textiles (E-Textiles 2021), Manchester, UK, 4 November 2021.

Abstract: The current study assessed the validity of a Kinematic Knee Sleeve (KiTT) against a gold-standard motion-capture system (Vicon, Oxofrd, UK). The relative knee angle, measured in the sagittal plane (RKA), was measured across a range of sporting movements to allow for comparisons and agreement between systems. The results demonstrate a high degree of validity of KiTT during a squat, deadlift, and leg curl, with partial validity of a leg extension (0.98, 0.97, 1.01, 1.31, respectively). KiTT serves as a valid method to collect information on the RKA. The KiTT appears to serve as a practical alternative to Vicon without sacrificing the quality of the data.

Keywords: kinematic; validity; knee; textiles; wearable; sensor; three-dimensional

1. Introduction

Three-dimensional (3D) motion-capture systems are acknowledged as the gold-standard technological method to investigate movement patterns and biomechanics [1]. Motion-caputre sysyems are capable of recording joint angles such as the relative knee angle in the sagittal plane (RKA), the displacement of segments, and the angular motion of joints/segments [2,3]. RKA is commonly measured when assessing squat depth to provide the user/coaches with information relating to range of motion or strength improvements that can be used to develop effective strength and condtioning strategies and rehabilitation plans [1–4]. However, when recording motion through such systems, real-time data are not available, comprimising the data's value during a particular session. Additionally, these systems are not as accessbile and are coupled with the need for specialist equipment and training. Specialist moition-capture systems are also fixed within a specific location, affecting ecological validity.

Wearable sensors that can be worn away from specilist settings and provide real-time and instantaneous data to users and coaches are an attractive propostion [5]. Being able to access real-time data allows exercise or training methods to be adjusted instantly, suiting the needs of the session to aid performance/rehabilitative progress in a way that is not possible with fixed and speicalist motion-capture systems [5,6]. Previous wearable sensors, such as smart watches, have focused on comfort for the user, rather than the quality of data. As a result, accuracy in the data is often lost, leading to unrelaible and invalid data, limiting its use and applicability within applied practice [7,8].

An innovative technology developed by Footfalls and Heartbeats (UK) Limited (Nottingham; FHL) aims to bridge the gap between comfort and the production of valid data sets. The Kinematic Knee Sleeve (KiTT) is a custom-knitted smart wearable knee sleeve, which is the first of its kind that knits the sensor directly into the fabric. Part of the KiTT is an electronics module, allowing data from the textile strain sensor to be transmitted to a portable device; however, validation against criterion methods has yet to be conducted.

Accordingly, the aim of the current study was to investigate the validity of the KiTT against the gold-standard motion-analysis system.

2. Materials and Methods

2.1. KiTT Structure

KiTT (Version 7.3) was knitted as a single piece of textile using a Stoll CMS ADF 32 W knitting machine (Karl Mayer Stoll, Reutlingen, Germany). The main body of KiTT consists of lycra (22 dtex) and polyamide 6.6 (78/24/1 dtex, Zimmerman, Weiler-Simmerberg, Germany). This combination of yarn was plated with nylon (78/1 dtex, Progressive Threads Ltd., Nottingham, UK). The cuff only consists of lycra (78/20/1 dtex, Stretchline, Nottingham, UK).

The textile strain sensor was measured 85mm $\times$ 7mm (height $\times$ width), consisting of silver-plated nylon yarn (Statex Shieldex®, 117/17 dtex; electrical resistivity <1.5 KΩ/m, Bremen, Germany), which was knitted alongside regular nylon yarn (78/1 dtex, Progressive Threads Ltd., Nottingham, UK). The transmission lines consisted of silver-plated nylon yarn (Statex Shieldex®, 235/36 dtex; electrical resistivity <80 KΩ/m, Bremen. Germany).

2.2. Participants

Following informed consent, 10 participants (8 male) were recruited for the current study, with an average age of 30.1 $\pm$ 11.7 years, weight of 78.5 $\pm$ 15.7 kg, and height of 177.7 $\pm$ 8.4 cm. Ethics approval was provided by the University of Derby Human Science Research Ethics Committee (ETH2021-0579). The inclusion criteria ensured that participants had >2 years experience of resistance training; completed ~150 min/week of moderate-intensity exercise; and completed the University health screen questionnaire.

2.3. Procedure and Protocol

Two data-collection systems were used for the current study; KiTT (version 7.3, *FHL*) and a Vicon motion-capture system (Oxford, UK). KiTT requires a small electronic unit to be connected to press-studs within the knee sleeve. This allows data to be transferred concurrently to a base station connected to a Windows 10 PC during the sporting movement. KiTT was worn on the left knee with the electronics lateral to the knee. Vicon has been used extensively in the assessment of human movement research [1]. Vicon utilises a Vicon Vantage Capture System, along with two Vicon 720p Colour Bonita Cameras. A total of 16 retro-reflective markers were attached to the participant following the Plug-In GAIT Lower-Body AI model created by the Vicon Nexus system.

The current study used four sporting exercises. Back squat (SQ) and traditional deadlift (DL) were weighted with a self-selected weight (33.5 $\pm$ 12.3kg) and was not sufficient enough to induce fatigue throughout the study. The remaining exercises, a leg curl (LC) and a leg extension (LE), were modified due to machine availability. A resistance band, ankle cuff, and dumbbell were configured to allow for either knee flexion (LC) or extension (LE).

Participants attended the Human Performance Unit at the University of Derby on three occasions. The first session included a full familiarisation, and sessions two and three comprised of experimental data collection. Data-collection sessions used only one motion-capture system, which was randomised for each participant's visit. Participants completed all exercises in a set order, with each exercise consisting of five repetitions followed by 2 min rest. The final visit was identical in design but involved a second motion-capture system.

2.4. Data Analysis

RKA was only recorded from the left limb due to KiTT only being able to record left-limb motion data. From Vicon's data output, only left-limb RKA was taken for the current study. KiTT's raw data were converted into relative knee angle through hysteresis analysis and reference points for the whole movement [9]. Vicon quantifies RKA through the automatic tracking of retro-reflecive markers and embedded formulas and equations [1].

Mean and standard deviations were calculated, enabling comparisons between the two motion-capture systems. Bland and Altman (B&A) plots were used to assess the degree of validity of KiTT when compared to criterion methods. A 95% confidence interval was used to identify agreement within the two systems [10]. A peak angle, defiend as the point of greatest knee flexion (SQ, DL, and LC) or greatest knee extension (LE), was when RKA was recorded.

3. Results

Raw data display high validity between the two motion-capture systems (Table 1). Small–moderate differences (2–26%) were displayed in the KiTT compared to Vicon across each exercise (2.69° SQ, 1.51° DL, 1.48° LC, and 2.85° LE). There was high validity across three of the four exercises in the KiTT compared to Vicon, which can be observed in Figure 1.

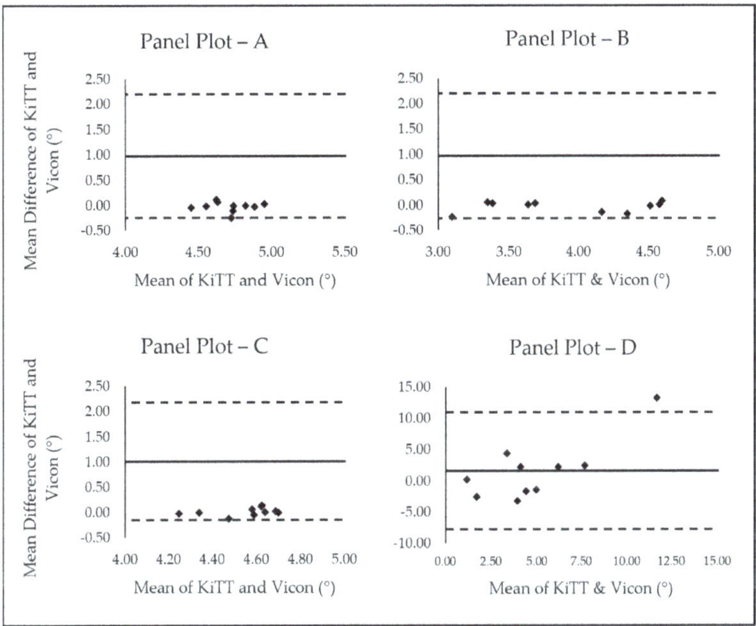

Figure 1. Panel plot of four Bland and Altman Plots: (**A**) B&A plot of the squat; (**B**) B&A plot of the deadlift; (**C**) B&A plot of the leg curl; (**D**) B&A plot of the leg extension.

Table 1. Relative knee angle from each sporting exercise (*n* = 10).

Collection System	Relative Knee Angle (°)			
	Squat	Deadlift	Leg Curl	Leg Extension
KiTT	111.33 ± 17.62	58.16 ± 31.06	96.44 ± 15.08	6.72 ± 6.02
Vicon	114.02 ± 18.02	59.67 ± 30.76	94.96 ± 12.52	3.87 ± 2.15
Similarity (%)	98	97	98	74

Before creating B&A plots, a test for heteroscedasticity was conducted. A positive result was found with SQ, DL, and LC, but not LE. As a result of this, SQ, DL, and LC raw data were translated into natural logarithmic data for the panel plots. Panel plots A, B, and C display the data points within the 95% Limits of Agreement, establishing validity in the KiTT. Panel plot D displays 90% of the data fitting within the 95% Limits of Agreement, establishing only partial validity with the KiTT for this specific exercise.

4. Discussion

The KiTT is a wearable technology that demonstrates a high level of validity against criterion laboratory assessment methods when completing whole body-exercise used in strength and rehabilitation environments. This is important as the KiTT could capture data in a non-controlled and specialist environment that is accessible by wider user groups. Motion-capture systems require markers to be affixed to the user, along with the calibration of the camera system and time-consuming data processing [1]. The KiTT only requires the sleeve to be worn on the user, with a Bluetooth connection established to the base station. For users and coaches, this can be invaluable, as data-collection time is significantly reduced, with data being generated instantaneously. Unlike motion-capture systems, the KiTT is not bound to a performance area where movement may be restricted; leading to increased ecological validity.

Low-similarity measures of RKA during the LE can be explained by the timing of measurements. The peak knee angle during the LE occurs when the textile is under rebound, where there is no tension throughout the sensor. As a result of this, it was not possible to obtain an accurate reading of RKA for this specific exercise, unlike SQ, DL, and LC. Through research and development, this issue may be resolved, leading to consistently valid data throughout collection.

Fixed motion-capture systems require direct line-of-sight to the markers, which can make the system inaccessible for specific exercises, such as squats and leg curls. On the other hand, the KiTT provides a unique method of assessment that is not bound to a specific performance area. Therefore, future research into the KiTT's validity and practicality should consider more dynamic and explosive movements. As the scope of this study was only inclusive of movements performed in a controlled environment, there is little/no evidence that the KiTT is a valid measurement tool in a more practical environment, thus warranting more extensive investigation into its practical uses and ecological validity.

Author Contributions: Conceptualization, funding acquisition, and resources, S.M. and M.F.; methodology, writing—original draft preparation, and writing—review and editing, N.T. T.O. and M.F.; software, S.M. and T.O.; validation and formal analysis, N.T. and M.F.; and supervision and project administration, S.M. T.O. and M.F. All authors have read and agreed to the published version of the manuscript.

Funding: This research received no external funding.

Institutional Review Board Statement: This study was conducted according to the guidelines of the Declaration of Helsinki and approved by the supervisors of the current study and the Ethics Committee of the University of Derby (Project Code: ETH2021-0579 on 5 March 2021).

Informed Consent Statement: Informed consent was obtained from all subjects involved in the study.

Data Availability Statement: The data presented in this study are available on request from the corresponding author. The data are not publicly available due to ethical obligations.

Acknowledgments: The authors gratefully recognise the work of Ruth Ashton.

Conflicts of Interest: The authors declare no conflict of interest.

References

1. van der Kruk, E.; Reijne, M.M. Accuracy of human motion capture systems for sport applications; state-of-the-art review. *Eur. J. Sport Sci.* **2018**, *18*, 806–819. [CrossRef] [PubMed]
2. Schurr, S.A.; Marshall, A.N.; Resch, J.E.; Saliba, S.A. Two-dimensional video analysis is comparable to 3D motion capture in lower extremity movement assessment. *Int. J. Sports Phys. Ther.* **2017**, *12*, 163. [PubMed]
3. Favre, J.; Jolles, B.; Aissaoui, R.; Aminian, K. Ambulatory measurement of 3D knee joint angle. *J. Biomech.* **2008**, *41*, 1029–1035. [CrossRef] [PubMed]
4. Morgan, D.L.; Proske, U. Popping sarcomere hypothesis explains stretch induced muscle damage. *Clin. Exp. Pharmacol. Physiol.* **2004**, *31*, 541–545. [CrossRef] [PubMed]

5. Schneider, C.; Hanakam, F.; Wiewelhove, T.; Döweling, A.; Kellmann, M.; Meyer, T.; Pfeiffer, M.; Ferrauti, A. Heart Rate Monitoring in Team Sports-A Conceptual Framework for Contextualizing Heart Rate Measures for Training and Recovery Prescription. *Front. Physiol.* **2018**, *9*, 639. [CrossRef] [PubMed]
6. Yavuz, H.U.; Erdag, D. Kinematic and Electromyographic Activity Changes during Back Squat with Submaximal and Maximal Loading. *Appl. Bionics Biomech.* **2017**, *2017*, 9084725. [CrossRef] [PubMed]
7. Aroganam, G.; Manivannan, N.; Harrison, D. Review on Wearable Technology Sensors Used in Consumer Sport Applications. *Sensors* **2019**, *19*, 1983. [CrossRef] [PubMed]
8. Zhao, Y.; You, Y. Design and data analysis of wearable sports posture measurement system based on Internet of Things. *Alex. Eng. J.* **2021**, *60*, 691–701. [CrossRef]
9. Kubo, K.; Kanehisa, H.; Fukunaga, T. Effects of resistance and stretching training programmes on the viscoelastic properties of human tendon structures in vivo. *J. Physiol.* **2002**, *538 Pt 1*, 219–226. [CrossRef] [PubMed]
10. Bland, J.M.; Altman, D.G. Statistical methods for assessing agreement between two methods of clinical measurement. *Lancet* **1986**, *1*, 307–310. [CrossRef]

engineering proceedings

MDPI

Proceeding Paper

Design of Textile Antenna for Moisture Sensing †

Irfan Ullah *, Mahmoud Wagih and Steve P. Beeby

School of Electronics and Computer Science, University of Southampton, Southampton SO17 1BJ, UK; mahmoud.wagih.mohamed@soton.ac.uk (M.W.); spb@soton.ac.uk (S.P.B.)
* Correspondence: i.ullah@soton.ac.uk
† Presented at the 3rd International Conference on the Challenges, Opportunities, Innovations and Applications in Electronic Textiles (E-Textiles 2021), Manchester, UK, 3–4 November 2021.

Abstract: This study reports a design of an e-textile microstrip patch antenna for wireless sensing of the moisture content of a fabric substrate. The microstrip patch antenna with a proximity coupled feeding line is implemented on two layers of polyester felt substrate. The performance of the antennas in terms of the reflection coefficient S11 is measured, indicating that the resonance frequency of the antenna shifts to a lower frequency for moisture contents ranging from 20% to 100%. This is the result of a change in the dielectric constant and the loss tangent of the substrate material caused by the presence of moisture. The proposed moisture sensor exhibits high linearity and higher sensitivity than state-of-the-art textile-based antenna sensors, and is suitable for a variety of applications such as sweat and wound monitoring.

Keywords: antenna sensor; chipless moisture sensor; resonant frequency

1. Introduction

There has been an increased interest in recent years in designing antenna-based sensors, due to their simple structure, low-cost, battery-free, and wire-free operation [1]. The majority of wireless sensors use the resonant technique to sense a variety of materials from liquid characterisation to temperature sensing, and even crack and strain monitoring. Moisture causes the antenna to detune and introduces losses into the textile antenna. As water has a much higher and more stable dielectric constant, it can significantly change the dielectric properties of the fabric leading to a shift in the antenna resonance frequency. By monitoring the resonant frequency of the antenna, which changes as a function of the water content absorbed by the fabric substrate, a wireless sensor can be realized. Such a moisture sensor can be used for various applications such as for measuring water drops or sweat and fluid loss in wound care.

In Refs. [2,3], a planar inverted F-antenna (PIFA) was implemented by embroidering conductive yarn on a denim textile for moisture sensing. However, non-linearity in the antenna response meant that the sensor did not provide accurate results. Furthermore, the patch was shorted at the end, which required the destruction of the textile substrate. As a result, the antenna was hard to fabricate on textile materials.

This study investigates the performance of microstrip patch antennas in terms of variation in the reflection coefficient in response to different moisture contents. The microstrip patch antennas with proximity coupled feeding lines were implemented in two-layer felt substrates, and the dimensions of the antenna were tuned to resonate at 2.45 GHz. The antennas were tested with different moisture contents and the results showed that the resonance frequency of the antenna shifted to a lower frequency, demonstrating its suitability to be used as a moisture sensor. Compared with alternative textile antenna designs, the proposed antenna designs are more suitable for integration within textiles, and can be easily fabricated.

check for
updates

Citation: Ullah, I.; Wagih, M.; Beeby, S.P. Design of Textile Antenna for Moisture Sensing. *Eng. Proc.* **2022**, *15*, 11. https://doi.org/10.3390/engproc2022015011

Academic Editors: Kai Yang and Russel Torah

Published: 14 March 2022

Publisher's Note: MDPI stays neutral with regard to jurisdictional claims in published maps and institutional affiliations.

2. Antenna Design

The geometry of the proximity coupled microstrip patch antenna is shown in Figure 1a. The length and width of the patch are 52 mm and 46 mm, respectively. The patch length can be optimised to tune the antenna's resonance frequency, i.e., increasing the length of the patch will shift the resonance frequency to a lower band. A 50 Ω microstrip feed line, 4.15 mm wide and 31 mm long, was designed and symmetrically positioned under the patch on top of substrate 2. Moreover, a 1-mm-thick polyester substrate with a dielectric constant of (ε_r) of 1.2 and loss tangent (δ) of 0.02 at 2.45 GHz was used to simulate the antenna in CST Microwave Studio [4] in two configurations: (i) polyimide laminates [5], and (ii) copper fabric [6], used for the patch, the microstrip, and the ground plane.

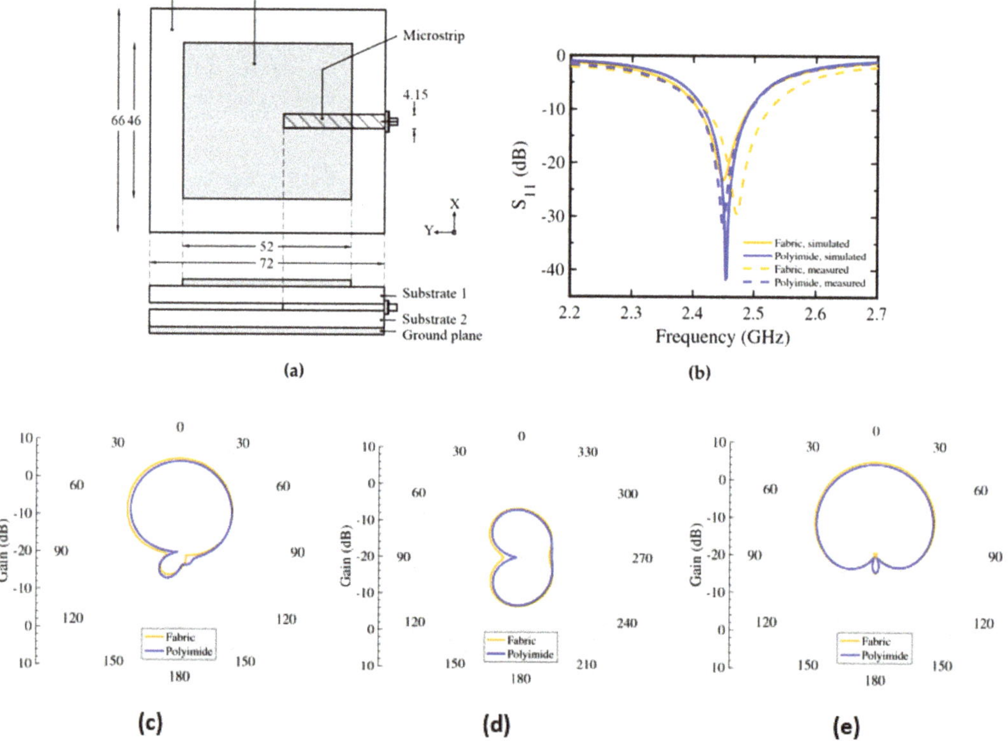

Figure 1. (**a**) Geometry of the proximity coupled microstrip patch antenna; (**b**) simulated and measured reflection coefficient S_{11} of the polyimide and fabric antenna designs; antenna gain in the (**c**) yz–plane, (**d**) xy–plane and (**e**) xz–plane.

The simulated and measured reflection coefficients of the antennas are shown in Figure 1b, indicating that both antennas were well matched (S11 > –10 dB) at 2.45 GHz. There was a slight shift in the measured frequency response of the fabric antenna, which was caused by cutting the patch a little short. Figure 1c–e shows the radiation pattern of the patch in yz–plane, xy-plane and xz-plane, respectively, and it can be observed that both antennas had a gain of about 4.98 dB with a radiation efficiency of 49.5% at 2.45 GHz.

3. Measurements and Results

The antenna (Figure 1a) was fabricated. In the first prototype, a copper laminate was used for the patch, the microstrip, and the ground plane, while a copper fabric was used in the second prototype, realizing a fully textile, breathable, and flexible antenna, as shown in Figure 2a. A repositionable adhesive pray [7] was used to attach the coppers to the substrates. The antenna designs (Figure 2a) were baked in the oven for about 5 hours at 105 °C to dehumidify the substrate, and then the reflection coefficient measurements were performed with a calibrated network analyser, as shown in Figure 2b. Figure 2c,d shows a shift in the resonance frequency as the antenna substrate gradually absorbs more water, where 0% refers to the dry antenna. The amount of the water content absorbed by the fabric was calculated as:

$$Moisture\ content\ (\%) = \frac{m_{wet} - m_{dry}}{m_{dry}} \times 100 \qquad (1)$$

where m_{wet} and m_{dry} are referring to the wet and dry weight of the antenna. The results show that the resonance frequency of both antennas shifted to a lower frequency when the different moisture content was applied. It can be observed that the maximum water content for the fabric antenna reached 77%. This was due to the amount of water lost from the substrate through the ground plane, as the fabric was not water-resistant compared to the copper antenna. At the dry state, both antennas resonated at the 2.45 GHz ISM band. On the other hand, when the antennas were fully wet, the antenna made of polyimide resonated at 1.34 GHz, while the antenna made of ohmic sheet resonated at 1.45 GHz. The shifting in the resonance frequency of the antennas determined the moisture content absorbed by the antenna substrate, and hence the sensor response. Comparing with alternative textile antennas, the proposed patch antenna is more suitable for textile applications as it has a more linear response and can be easily fabricated. The results indicated that the fully textile antenna had a higher sensitivity value of 0.532% compared to the copper antenna and the antenna proposed in Refs. [2,3], which had sensitivity values of 0.443% and 0.463%, respectively.

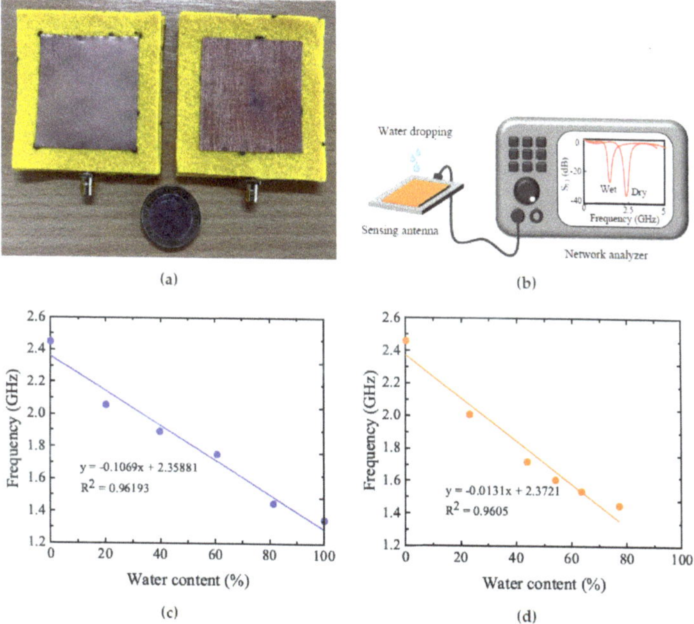

Figure 2. (**a**) Fabricated antennas; (**b**) measurement setup; (**c**) resonance frequency against water content for polyimide antenna; and (**d**) resonance frequency against water content for fabric antenna.

4. Conclusions

Two patch antennas operating at 2.45 GHz have been demonstrated in this paper, enabling low-cost moisture sensing. The resonance frequency of the antenna shifts to a lower frequency band as the antenna absorbs more moisture. This shift in the resonance frequency of the antenna indicates the amount of moisture absorbed by the fabric. It was observed that the full fabric patch antenna design was more sensitive compared to the polyimide antenna. The antennas are flexible, low-cost, easy to fabricate, and can be seamlessly integrated with clothing for on-body applications. The objective of our future work is to investigate different types of fabrics which are more absorbent, such as escalade, cotton and linen.

Author Contributions: Conceptualization, I.U. and M.W.; methodology, I.U.; software, I.U.; validation, I.U., M.W. and S.P.B.; formal analysis, I.U.; investigation, I.U.; writing—original draft preparation, I.U.; writing—review and editing, M.W.; supervision, S.P.B.; project administration, S.P.B.; funding acquisition, S.P.B. All authors have read and agreed to the published version of the manuscript.

Funding: This research was funded by European Regional Development Fund (ERDF) via its Interreg V France (Channel) England programme: Smart Textile for Regional Industry and Smart Specialisation Sectors (SmartT). The work of Steve Beeby was supported by the Royal Academy of Engineering under the Chairs in Emerging Technologies Scheme.

Institutional Review Board Statement: Not applicable.

Informed Consent Statement: Not applicable.

Data Availability Statement: Not applicable.

Conflicts of Interest: The authors declare no conflict of interest.

References

1. El Gharbi, M.; Fernández-García, R.; Ahyoud, S.; Gil, I. A review of flexible wearable antenna sensors: design, fabrication methods, and applications. *Materials* **2020**, *13*, 3781. [CrossRef] [PubMed]
2. Bonefačić, D.; Bartolic, J. Embroidered textile antennas: influence of moisture in communication and sensor applications. *Sensors* **2021**, *21*, 3988. [CrossRef] [PubMed]
3. Bonefačić, D. Textile antenna as moisture sensor. In Proceedings of the 14th European Conference on Antennas and Propagation (EuCAP), Copenhagen, Denmark, 15–20 March 2020; pp. 1–3.
4. CST. Microwave Suite 2017. Available online: https://www.cst.com (accessed on 22 November 2021).
5. GTS. GTS 7800–Copper Polyimide Laminates. Available online: https://www.gtsflexible.com/ (accessed on 22 November 2021).
6. Meftex. Meftex 30. Available online: https://www.meftex.cz (accessed on 22 November 2021).
7. 3M. 3M Spray Mount. Available online: https://www.3m.co.uk/3M/en_GB/p/d/v000143885/ (accessed on 23 November 2021).

Proceeding Paper

Image Detection and Responsivity Analysis of Embroidered Fabric Markers Using Augmented Reality Technology [†]

Anuja Pathak, Ian Mills and Frances Cleary *

Mobile Ecosystem and Pervasive Sensing (MEPS), Walton Institute, X91 P20H Waterford, Ireland; anuja.pathak@waltoninstitute.ie (A.P.); ian.mills@waltoninstitute.ie (I.M.)
* Correspondence: frances.cleary@waltoninstitute.ie; Tel.: +353-51-302-920
† Presented at the 3rd International Conference on the Challenges, Opportunities, Innovations and Applications in Electronic Textiles (E-Textiles 2021), Manchester, UK, 3–4 November 2021.

Abstract: In this paper, we investigate the use of augmented reality technology within an E-textile environment. We place particular emphasis on the analysis of key performance and responsiveness metrics when utilizing augmented reality (AR) applications for embroidery-based logo/design image detection and recognition. To support this analysis and validation, we designed and created four test embroidered images, a fabric quilt with embroidered marker images, and a supporting augmented reality application. From an E-textile point of view, we explore the effects of high/low contrast thread colors, diverse light levels (lux measurements), and the range of angles at which the mobile device/camera, with the associated AR application, can be pointed towards the fabric-embroidered marker. This allows us to assess the level of functionality and responsiveness of the AR application and the overall performance in the testing environment, enabling more fluid usability of the AR-enabled E-textile application.

Keywords: augmented reality application; E-textiles; responsiveness; embroidered images

check for
updates

Citation: Pathak, A.; Mills, I.; Cleary, F. Image Detection and Responsivity Analysis of Embroidered Fabric Markers Using Augmented Reality Technology. *Eng. Proc.* **2022**, *15*, 12. https://doi.org/10.3390/engproc2022015012

Academic Editors: Steve Beeby, Kai Yang and Russel Torah

Published: 18 March 2022

Publisher's Note: MDPI stays neutral with regard to jurisdictional claims in published maps and institutional affiliations.

1. Introduction

In this paper, an AR application was developed to specifically detect and recognize embroidery-based markers in a fabric environment. Each Embroidered marker of varying colors were created, and the thread design of each marker images were selected in order to be able to compare the effects of high- and low-contrast thread colors, as shown in Figure 1a,b, and their detection responsivity by the AR application. Several observations were made, as shown in Tables 1 and 2, Also each marker was assigned an animated 3D model which was displayed on a mobile/tablet screen, as shown in Figure 2a. We further investigated and explored the effects of various other factors affecting the response time of the application, such as recognition accuracy at seven different angles and at varying light levels. These observations allowed us to identify the most suitable ranges for the above-mentioned factors and enabled us to effectively enhance the response time of the application, hence improving the interaction time for users and making the AR textiles application more user-friendly and data-driven.

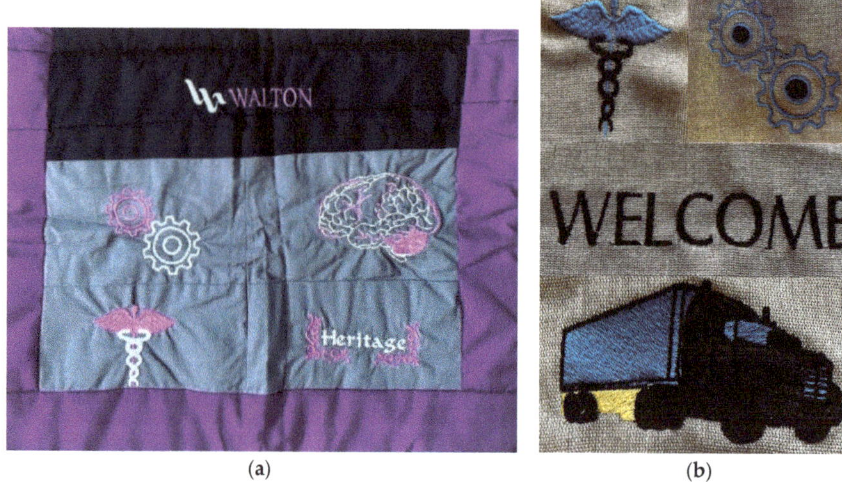

(a)   (b)

Figure 1. (**a**) The left image is the embroidery-woven wall hanging (high-contrast colors) with different embroidery symbols, which represent different animated 3D models. These 3D assets are triggered when the above-shown markers are recognized by the augmented reality camera. The above-shown markers are embridered in high-contrast colours and on a gray/black background, which make them easy to detect by an AR camera. (**b**) The right image is the embroidery-woven wall hanging (low-contrast colors) with different embroidery symbols. The above-embroidered symbols are mostly blue and black in color on a light, jute-textured background, which makes them difficult to recognize by an AR camera.

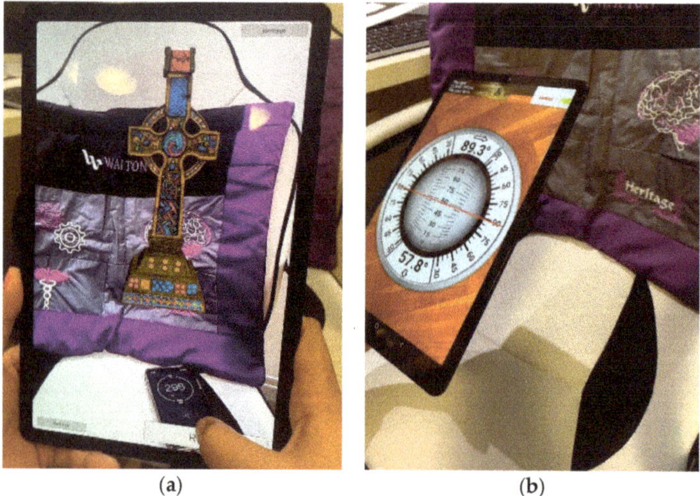

(a)   (b)

Figure 2. (**a**) The left figure is the AR view on a tablet. A light (lux) meter application on a different device is used to detect the amount of light. We can see in the image that a 3D object is triggered because of an embroidery marker on the wall hanging. (**b**) The right figure shown above is the apparatus used to measure angle using a rotating sphere inclinometer application. We can see that the device is inclined at a 60° angle and the inclinometer detects the angle.

Table 1. The table below contains the observations made for the response time at seven different angles at which the device is held for high- and low-contrast markers. Here, we can see that the application responds fastest when the phone is between 75° to 105° for both low- and high-contrast markers.

Sr. No.	Angle of the Device Held (x-Axis)	Response Time for High Contrast (s) (y-Axis)	Response Time for Low Contrast (s) (y-Axis)
1.	45°	40.28 s	52.27 s
2.	60°	30.28 s	38.25 s
3.	75°	16.17 s	16.11 s
4.	90°	5.43 s	7.51 s
5.	105°	8.68 s	8.39 s
6.	115°	15.32 s	26.32 s
7.	120°	31.12 s	43.14 s

Table 2. The table below contains the observations made for the response time at fifteen different light levels on high- and low-contrast markers. Here we can see that the fastest response of the application is when the light levels are moderate (232 lux -141 lux) for both high- and low-contrast markers.

Sr. No.	Brightness Level (lux) (x-Axis)	Response Time for High Contrast (s) (y-Axis)	Response time for Low Contrast (s) (y-Axis)
1.	337	26.66	45.07
2.	325	20.12	37.59
3.	318	17.35	30.35
4.	312	15.03	28.03
5.	308	13.77	28.77
6.	258	11.23	26.23
7.	232	9.64	19.41
8.	201	5.99	18.39
9.	192	5.64	17.28
10.	172	5.02	18.98
11.	168	5.23	20.33
12.	141	16.29	27.65
13.	113	18.63	28.42
14.	46	28.47	32.26
15.	5	31.28	48.52

As per the observations recorded below, in Table 2 and Figure 3, we proved that the AR application takes more time to respond to embroidery-based markers if the light levels (lux measurements) are too high or too low. We can also prove that the response time of the application is enhanced when the light levels are moderate. The contrast levels of the threads used for the embroidery marker design have a major impact on the performance of the application. From the results, we see that the response time for the low-contrast markers is significantly higher than the response time for the high-contrast markers. Thus, we can say, from the observations shown in Tables 1 and 2, that the recognition operates at peak performance when we have a higher degree of contrast in the color of the threads. We used white and purple threads on a black and grey background for high-contrast markers, and blue and black threads for low-contrast markers. Observations of the response time of the

AR application when the device is held at different angles for both high- and low-contrast embroidered markers can be viewed in Figure 4; here the graph clearly shows that the most suitable angle range for a device to be held for best results, for both low- and high-contrast markers, is 75° to 105°.

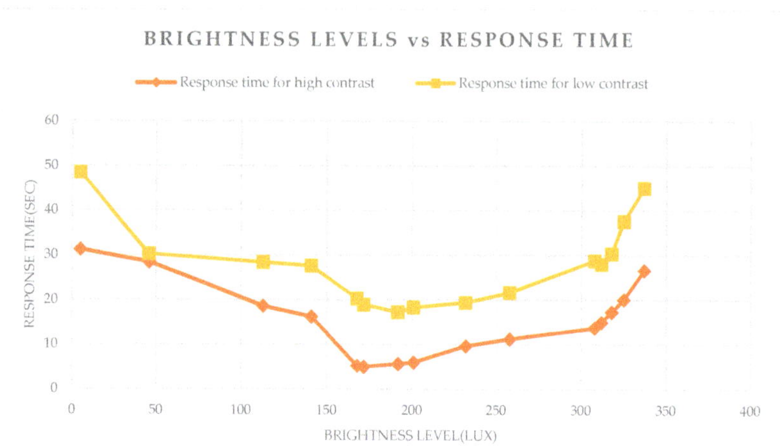

Figure 3. The above graph contains the observations made for the response time at fifteen different angles at which the device is held for high- and low-contrast markers.

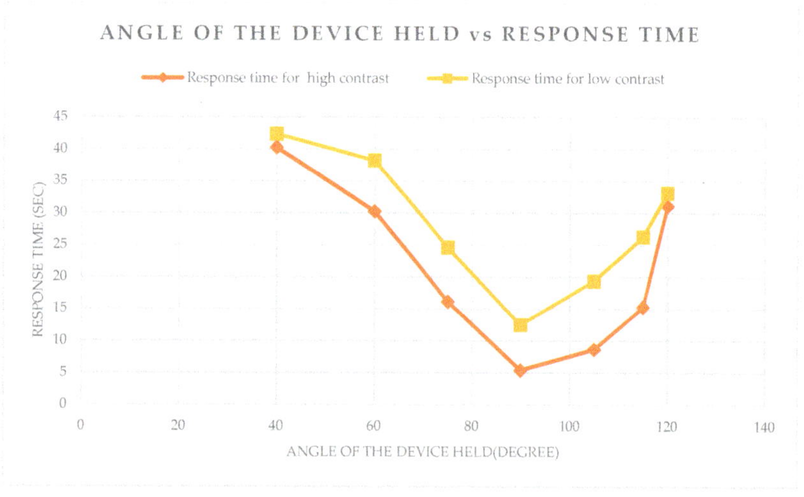

Figure 4. The above graph contains the observations made for the response time at seven different angles at which the device is held for high- and low-contrast markers.

2. Conclusions and Future Scope

In this paper, we explored the use of augmented reality technology within the E-textiles environment. We designed various experiments to find out how various factors—namely the brightness level, and the angle at which the device was held for both high- and low-contrast thread colors—affected the performance of the developed AR application.

In the future we would like to take this investigation further to see how the AR application would be affected if the embroidery markers were on curved surfaces with various light settings, thus simulating fabrics being worn by an individual. Doing so would allow us to further understand the nature of AR recognition for E-Textile use. We would also like to explore various color combinations and examine the effects of high-, medium- and low-contrast threads in more detail, to identify the optimum marker composition for E-Textile based embroidery. Our research will help us to build a more sustainable and reliable Augmented Reality E-textiles application and create various uses for it, such as a personalized story-telling AR application blanket for children, wall hangings, and many more. We also would like to explore the ways in which we can use augmented reality technology with E-textiles for health-monitoring purposes. Using a combination of infrared sensitive fabrics and the sensor array present in modern mobile devices, we aim to collect the captured data, process them using a custom shader to detect temperature changes in measured color metrics, and display the temperature in real time on the mobile device in augmented reality. Thus, we aim to deliver a low-cost health-detection feature, as well as a means to evaluate smart fabric solutions in an unobtrusive manner.

Author Contributions: Conceptualization, A.P., F.C. and I.M.; Methodology, A.P.; Software, A.P.; validation, I.M., F.C. and A.P.; Formal analysis A.P.; investigation, A.P.; resources, F.C.; data curation, A.P.; Writing—original draft preparation, A.P.; Writing—review and editing, I.M. and F.C.; Visualization A.P.; Supervision, I.M. and F.C.; Project administration, I.M. All authors have read and agreed to the published version of the manuscript.

Funding: This research received no external funding.

Institutional Review Board Statement: Not Applicable.

Informed Consent Statement: Not Applicable.

Data Availability Statement: https://owncloud.tssg.org/index.php/s/W5t3EB88MSjbwhU; https://owncloud.tssg.org/index.php/s/jgflPmE62R8LNoU.

Conflicts of Interest: The authors declare no conflict of interest.

engineering proceedings

MDPI

Proceeding Paper

5G-Enabled E-Textiles Based on a Low-Profile Millimeter-Wave Textile Antenna [†]

Mahmoud Wagih [1,*], Geoff S. Hilton [2], Alex S. Weddell [1] and Steve Beeby [1]

[1] School of Electronics and Computer Science, University of Southampton, Southampton SO17 1BJ, UK; asw@ecs.soton.ac.uk (A.S.W.); spb@ecs.soton.ac.uk (S.B.)
[2] Communications Systems and Networks Group, University of Bristol, Bristol BS8 1TH, UK; geoff.hilton@bristol.ac.uk
* Correspondence: mahm1g15@ecs.soton.ac.uk
[†] Presented at the 3rd International Conference on the Challenges, Opportunities, Innovations and Applications in Electronic Textiles (E-Textiles 2021), Manchester, UK, 3–4 November 2021.

Abstract: Wireless Body Area Networks (WBANs) are a key application underpinned by advances in electronic textiles (e-textiles). Achieving higher throughput, data-rate, network capacity, and delivering wireless power to miniaturized devices requires WBANs to operate at millimeter-wave 5G+ frequencies. This, however, imposes significant challenges on the antenna design to cope with the additional losses introduced by textile substrates. In this paper, the performance of a novel, high-efficiency, textile-based millimeter-wave antenna is investigated for wireless links with a wearable device. Indoor "real-world" channel gain measurements are used to evaluate the antenna's performance compared to anechoic gain measurements. Based on the measured channel gain between textile antennas, it is concluded that high-speed wireless links in the 24–30 GHz 5G+ spectrum could be realized with over one meter range using e-textile antennas.

Keywords: wearable sensor; antenna sensor; RF breathing sensor; wireless sensors; textile sensors; vital sign monitoring

Citation: Wagih, M.; Hilton, G.S.; Weddell, A.S.; Beeby, S. 5G-Enabled E-Textiles Based on a Low-Profile Millimeter-Wave Textile Antenna. *Eng. Proc.* **2022**, *15*, 13. https://doi.org/10.3390/engproc2022015013

Academic Editors: Russel Torah and Kai Yang

Published: 15 March 2022

Publisher's Note: MDPI stays neutral with regard to jurisdictional claims in published maps and institutional affiliations.

1. Introduction

Wireless Body Area Networks (WBANs) are widely regarded as the most scalable and convenient method of connecting wearable devices. On-body WBANs, where wearables communicate between each other on the same body, as well as off-body WBANs, where the wearables communicate with a gateway away from the body, have attracted significant research interest [1].

Recently adopted for 5G communications, millimeter-wave (mmWave) bands are expected to enable high data-rate communications as well as energy-efficient networks based on wireless power transmission (WPT) [2]. Nevertheless, at mmWave frequencies, antennas are typically fabricated on low-loss, dielectric-stable substrates to achieve high gain and radiation efficiency. Recently, a range of textile-based antennas have been proposed for mmWave wearable applications [3,4]. A range of e-textile fabrication methods, including the laser ablation of copper films [3], photolithography on polyimide flexible circuit filaments [4], as well as screen printing combined with heat transfer [5], have been used to realize mmWave components such as antennas, transmission lines, and capacitors on textile substrates. While the radiation efficiency of textile-based mmWave antennas remains below their non-textile counterparts, the antennas' gains, characterized through anechoic radiation pattern measurements, show the potential for high-speed communications and WPT. In most mmWave network architectures, directional antenna arrays with high gain are used [1]. Therefore, it is essential to experimentally characterize the link budget between e-textile, wearable mmWave antennas, where the highly dynamic and mobile body-centric environment may hinder effective communication between mmWave wearables.

In this paper, a state-of-the-art, textile-based mmWave microstrip antenna implemented on a woven polyester substrate is investigated for wearable applications in line of sight and non-line of sight cases. Through empirical channel gain measurements, it is shown that e-textile mmWave antennas are suitable for operation in this environment.

2. E-Textile mmWave Antenna Fabrication

The key requirements for maintaining a low conductor loss in mmWave bands are high conductivity and low roughness [3,5]. To explain, skin-depth-induced losses increase with frequency. Therefore, rough conductors are expected to introduce additional attenuation compared to those of smooth lines. This can be attributed to the low quality and homogeneity of the surface of the conductor, where most of the currents are confined in microwave and mmWave bands [5]. Therefore, to realize a high-efficiency antenna with minimal conductive losses, the photolithography of commercially available, flexible copper laminates is used to fabricate the antenna. The fabrication steps are summarized in Figure 1. The flexible copper laminates are based on a low-loss (mmWave dissipation factor < 0.01) polyimide film. The textile substrate used is woven polyester with a relative permittivity $\varepsilon r = 1.95$ and dissipation factor $\tan\delta = 0.026$. Following the patterning of the copper sheets to create the antenna's radiating top layer, the flexible polyimide antenna can be laminated onto different textile substrates, including rough fabrics unsuitable for direct printing [5], as shown in Figure 1e.

Figure 1. Fabrication process of e-textile mmWave antennas for 5G+ applications based on etched polyimide copper laminates: (**a**) 25 μm-thick copper-clad Kapton; (**b**) photoresist deposition; (**c**) UV exposure through a film mask; (**d**) antenna traces following etching; (**e**) antenna traces adhered to fabric.

The antenna used in this work is based on a modified microstrip patch antenna with an inset feed, designed to maintain a matched input impedance over two resonant transverse magnetic (TM) second-order modes, to achieve a wider beamwidth than a conventional first-order mode microstrip antenna [6]. Figure 2a shows the layout and dimensions of the antenna, optimized for the woven polyester substrate. The bottom layer of the antenna is composed of a conductive ground plane, which can be realized using low-cost, breathable conductive fabrics [2] or a printed silver layer [3]. Figure 2b,c are photographs of the fabricated prototype, demonstrating its flexibility.

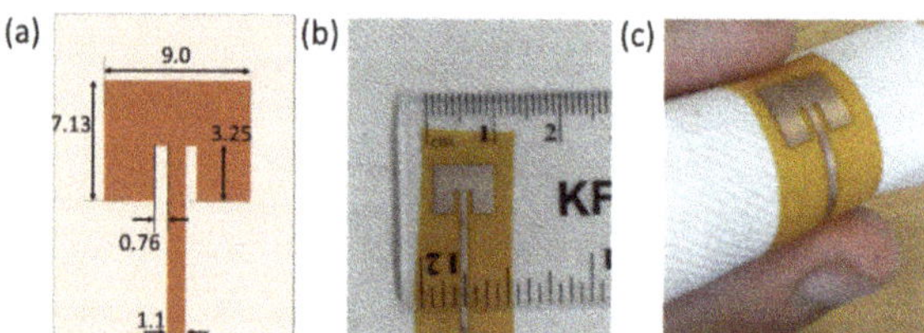

Figure 2. The Kapton-on-woven polyester mmWave antenna (**a**) layout and dimensions (in mm); (**b**) the fabricated device; (**c**) applied to a surface with approximate bend radius of 5 mm.

3. mmWave E-Textile Antenna Characterization

The fabricated prototype was characterized experimentally to evaluate its bandwidth and radiation properties. A key requirement of a mmWave antenna is the maintenance of a broad bandwidth [4]. To explain, a key advantage of operating at mmWave frequencies is the wider spectrum availability compared to sub-6 GHz bands [2]. For example, mmWave 5G uses the 26 and 28 GHz bands, with a combined bandwidth of over 4 GHz. To measure the antenna's response, a solder-terminated 1.85 mm connector was added to the microstrip feed. The antenna's reflection coefficient was measured using an Agilent E8361A PNA Vector Network Analyzer (VNA), calibrated using a standard electronic calibration kit. Figure 3 shows the measured reflection coefficient of the antenna as well as the simulated response, obtained through full-wave numerical electromagnetic simulation in CST Microwave Studio Suite.

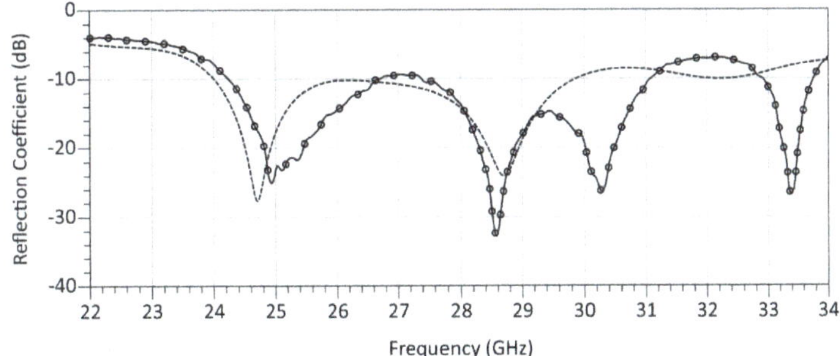

Figure 3. Simulated (no marker) and measured (solid with markers) reflection coefficient (S11) of the antenna, showing an S11 < −10 dB bandwidth from 24.2 to 31.1 GHz.

As observed in Figure 3, the antenna fulfills the bandwidth requirement of a 5G wearable antenna, where both the measured and simulated S11 responses indicate that over 90% of the power will be accepted by the antenna, demonstrating its suitability for integration in mmWave 24.2–30 GHz transceiver RF chains including amplifiers.

To evaluate the antenna's far-field properties and subsequent performance in real-world, on-body mmWave links, the antenna's radiation patterns were experimentally measured. The polarized radiation pattern measurements were performed relative to a standard 20 dBi gain waveguide horn antenna in an anechoic chamber. The total efficiency of the textile antenna (inclusive of impedance mismatch losses) was calculated relative to the horn antenna [4]. Figure 4a shows the measured 3D radiation patterns of the antenna, where it can be seen that the antenna maintains two main lobes, in line with a standard second-order mode microstrip patch antenna.

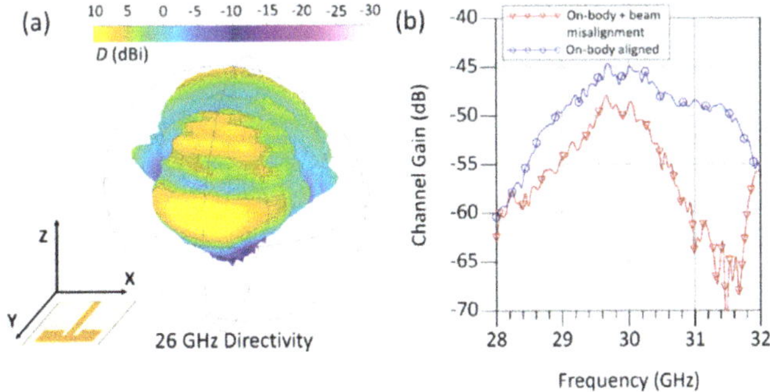

Figure 4. The far-field performance of the e-textile mmWave antenna: (**a**) measured 3D spherical radiation patterns; (**b**) measured channel gain between symmetric antennas on-body, with and without main beam alignment at a 60 cm distance.

The total efficiency of the antenna, with respect to the reference horn, was found to be around 60% ($\pm$5%). The antenna was experimentally characterized through channel gain measurements to show its suitability for on-body mmWave communications. In Figure 4b, it can be seen that over -55 dB forward transmission can be maintained between two symmetric textile antennas along the body with and without main beam alignment. Such a link budget would enable this e-textile antenna to integrate with recently reported, short-range, high-data-rate mmWave transceivers, such as low-power, high-speed backscattering modulators printed on flexible substrates [7].

4. Conclusions

In this paper, the performance of a mmWave e-textile antenna was characterized, demonstrating its suitability for short-range, high-speed communication in 5G mmWave bands. It is anticipated that low-profile and high-efficiency mmWave textile-based antennas, such as the microstrip patch considered in this study, will enable future WBANs to leverage the mmWave spectrum to improve their throughput and end-to-end efficiency.

Author Contributions: Conceptualization, M.W.; methodology, M.W.; investigation, M.W. and G.S.H.; writing—original draft preparation, M.W.; writing—review and editing, M.W., A.S.W. and S.B.; funding acquisition, S.B. All authors have read and agreed to the published version of the manuscript.

Funding: This work was supported by the UK Engineering and Physical Sciences Research Council (EPSRC) under Grant EP/P010164/1. The work of Steve Beeby was supported by the Royal Academy of Engineering under the Chairs in Emerging Technologies Scheme. Data access: The datasets used in this work are available from doi:10.5258/SOTON/D1931.

Institutional Review Board Statement: Ethical approval has been obtained from the FEPS Faculty Ethics Committee, University of Southampton, SO17 1BJ, application number: 64814, approved: 26 August 2021.

Informed Consent Statement: Not applicable.

Data Availability Statement: Not applicable.

Conflicts of Interest: The authors declare no conflict of interest.

References

1. Hall, P.; Hao, Y. *Antennas and Propagation for Body-Centric Wireless Communications*, 2nd ed.; Artech: Morristown, NJ, USA, 2012.
2. Rappaport, T.S.; Sun, S.; Mayzus, R.; Zhao, H.; Azar, Y.; Wang, K.; Wong, G.N.; Schulz, J.K.; Samimi, M.; Gutierrez, F. Millimeter Wave Mobile Communications for 5G Cellular: It Will Work! *IEEE Access* **2013**, *1*, 335–349. [CrossRef]
3. Chahat, N.; Zhadobov, M.; Muhammad, S.A.; le Coq, L.; Sauleau, R. 60-GHz Textile Antenna Array for Body-Centric Communications. *IEEE Trans. Antennas Propag.* **2013**, *61*, 1816–1824. [CrossRef]
4. Wagih, M.; Hilton, G.S.; Weddell, A.S.; Beeby, S. Broadband Millimeter-Wave Textile-Based Flexible Rectenna for Wearable Energy Harvesting. *IEEE Trans. Microw. Theory Tech.* **2020**, *68*, 4960–4972. [CrossRef]
5. Wagih, M.; Komolafe, A.; Hillier, N. Screen-Printable Flexible Textile-Based Ultra-Broadband Millimeter-Wave DC-Blocking Transmission Lines Based on Microstrip-Embedded Printed Capacitors. *IEEE J. Microw.* **2021**, *2*, 162–173. [CrossRef]
6. Wagih, M.; Hilton, G.S.; Weddell, A.S.; Beeby, S. Millimeter Wave Power Transmission for Compact and Large-Area Wearable IoT Devices based on a Higher-Order Mode Wearable Antenna. *IEEE Internet Things J.* **2021**. [CrossRef]
7. Kimionis, J.; Georgiadis, A.; Daskalakis, S.N.; Tentzeris, M.M. A printed millimetre-wave modulator and antenna array for backscatter communications at gigabit data rates. *Nat. Electron.* **2021**, *4*, 439–446. [CrossRef]

engineering proceedings

Proceeding Paper

Can Design for Disassembly Principles Inform Policy for E-Textiles Waste? †

Jessica Saunders

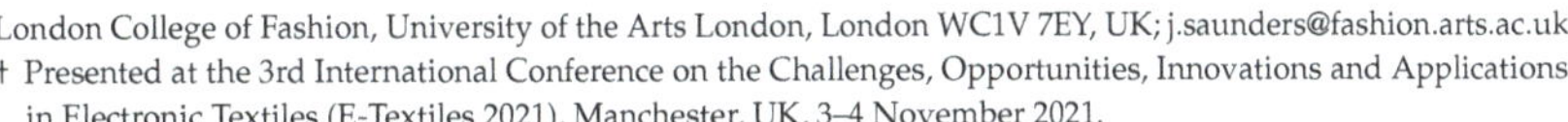

London College of Fashion, University of the Arts London, London WC1V 7EY, UK; j.saunders@fashion.arts.ac.uk
† Presented at the 3rd International Conference on the Challenges, Opportunities, Innovations and Applications in Electronic Textiles (E-Textiles 2021), Manchester, UK, 3–4 November 2021.

Abstract: The adoption of circular principles by the EU and UK have led to greater focus on waste streams and the recoverability of materials and components. This has translated into regulations such as WEEE for electronic waste. Textiles and nanomaterials lag behind with no definitive waste legislation. As e-textiles are generally made up of a combination of these three components, it means e-textile products end up in electrical recycling facilities where textile components are disposed of in landfill or incinerated together with embedded nanomaterials. Consultations with recycling facilities indicate product design is key in preparing for disassembly and recycling. By embedding design for disassembly thinking into the research and development of new e-textiles, this study aims to test whether e-waste policy can be informed by design for disassembly principles. The motivation for this research is to find an anticipatory legislative solution for future e-textiles waste.

Keywords: design for disassembly; Active Disassembly; e-textiles; recycle; waste policy; landfill

Citation: Saunders, J. Can Design for Disassembly Principles Inform Policy for E-Textiles Waste? *Eng. Proc.* **2022**, *15*, 14. https://doi.org/10.3390/engproc2022015014

Academic Editors: Russel Torah, Steve Beeby and Kai Yang

Published: 19 April 2022

Publisher's Note: MDPI stays neutral with regard to jurisdictional claims in published maps and institutional affiliations.

1. Introduction

The predicted monetary growth of wearable electronics by an additional USD 46 billion by 2025 [1], e-textiles waste is set to become a significant new waste stream. Planning for this eventuality is an opportunity to innovate in materials, products and recycling. To date, as the race to make flexible, wearable, functional and scalable e-textiles has intensified, little consideration has been given to the end-of-life implications of embedded electronics in our garments or medical devices. Nanomaterials are an integral part of new developments that seek to find ever more discreet ways to embed connectivity and function into e-textiles, bringing another element of complexity to an already problematic waste stream. There has been surprisingly little thought or research into how e-textiles can be disassembled, discarded or repurposed; furthermore, there has been little consideration into the use of raw materials, processing or disposal in line with European Circularity or Waste Electronic and Electrical Equipment (WEEE) directives. With Google [2] promising to embed our technology so that it "disappears" into the background, merging seamlessly into our lives and clothing in a world of "ambient computing", it appears that developments in the field are clearly well funded and moving at pace. Meanwhile, there is often a failure to mention sustainability, or end-of-life management.

2. Legislation and Regulation

The legislative landscape in relation to textiles, nanomaterials and electronics is varied. Textile waste is not officially regulated or managed with any enforceable standards globally (although the EU will bring in kerbside collection by 2025), relying on voluntary standards or regulations. Nanomaterials have no recovery or disposal policy, G20 Insights report (2017) [3] recommended member states "work together to develop efficient and unified policy efforts to regulate the field of nanotechnology, . . . (to) apply Precautionary Principle to all nanotechnology developments", also highlighting that the toxicity and chemical reactivity of nanoparticles are concerning. This means efforts and funding should be channelled into the future management of nanowaste [4]. A total of 27 EU countries, the UK and 41 others outside of the EU have national e-waste legislation/policy or regulation in place [5]; despite this, there is an ever-growing volume of electronics going to landfill with only 17% of electronic waste being recycled globally. It appears there is still a long way to go on waste management, as UK landfill textile waste is estimated to rise by 60% to 148 million tonnes by 2030 [6]. These facts provide an opportunity to bring new e-textiles to market with a known strategy and framework for disposal and recovery, which could lead the way for future policy relating to waste across electronics, textiles and nanomaterials. Current evidence from recycling facilities suggests the biggest barrier for e-textiles recovery is that the infrastructure does not exist to manage mixed-material waste. For example, when an electronics recycling facility receives e-textiles, they are not obligated or equipped to recycle the textile part of the product and will discard to landfill once any electrical parts, e.g., batteries, recoverable metals or plastics, are removed. Smart clothing waste is beginning to arrive in electronic recycling facilities, where recoverable electronic components can be removed and processed, leaving the textile portion discarded into landfill or incinerated.

Whilst considering legislation development that encourages innovation in e-textiles, it is worth considering previous debate and research on policy. As far back as 1800, policy makers and theorists such as Bentham [7], and more recently Marciniak [8], advocate preventative legislation, agreeing that repairing damage is worse than adopting antici-patory policy principles. The arguments against preventative legislation are strong, as a potential consequence is loss of individual and societal freedoms. However, there is differentiation and debate between prevention and the so-called 'precautionary principle', highlighted by Marciniak, who acknowledges that environmental politics have led to more precautionary directives or regulation. In relation to e-textiles and their development, the waste impact is speculative, but can be based on an understanding of waste habits in related fields such as electronics and textiles. It is safe to say that preventative measures prior to e-textile waste reaching unmanageable levels are advisable and would encourage innovation in disposal and reuse. Within the EU, precautionary principles have been adopted, not without debate on effective implementation. However, retrospectively, this approach may have avoided some previous environmental issues such as asbestosis and dichlorodiphenyltrichloroethane (DDT) [9].

Legislation and regulation are closely linked, but according to Kosti et al. 2019 [10], there is little academic discourse on the subject. "Legislation and regulation increasingly impact our lives", and interplay between the two can make a difference as to how effec-tive policy can be. Often regulation or standards are voluntary, relying on industry to comply. Policy makers have made use of voluntary standards and regulations as they are often designed by industry to bring stakeholders on board, particularly in environmental areas [11]. Records on textile and e-waste do not reflect well on the dynamic between voluntary regulation, standards and policy, though that is not to say the system cannot work in future.

3. Design for Disassembly

Dr Joseph Chiodo is the instigator of research into "Active Disassembly" [12] which was recognised by the Ellen McArthur Foundation [13] report which attracted academic and political attention. The report makes the case for a circular-based fashion system, stating that there is a need to "align clothing design and recycling processes". Both advocate the design phase as key to tackling waste by enabling circularity through reuse and disassembly. Chiodo's research in car manufacturing has influenced the industry to embed disassembly in design and manufacture, bringing economic gains in the process.

WEEE already contains a requirement to design for recovery. However, the main focus is on how to manage products at end of life, while the "design" part is often lost with little or no reporting, despite the added expectations of extended user responsibility (EPR) schemes. "Few or no quantitative targets or indicators on eco-design and waste prevention have been developed within EPR schemes, as all of them are designed around main objectives on waste collection and recycling [14]". Manufacturers are required to build strategies for disassembly into the design of their products. In the past, designing products such as cars rarely involved consideration of disposal, although some companies, such as BMW, have been proactive in this respect.

Many researchers have recognized the need for design to be central to waste issues, Preston, F. (2012) [15] suggests "improving product design, (and) improving capture of products at end of life". Goldsworthy, K. and Ellams, D. (2019) [16] demonstrate that design should be considered an important part of a circular economy model by creating a "unique opportunity to develop a systematic, life cycle stakeholder approach from which to explore 'what designers need to know when designing for circularity". Bocken, N. M. P. et al. (2016) [17] echo Goldsworthy in their recommendation that circular considerations are taken into account during the early speculative design phase of product development. Bocken highlights Circular Design and Circular Economics require a combination of science, business and design to be in synergy.

Design for disassembly principles, therefore, are a key part of influencing circular aspirations and an important factor in potentially aiding waste management. Moreno, M. et al. (2016) [18] advocate that designer's should be "solution providers" rather than "object creators" and the 2016–2020 ENTeR project [19] identified missing elements in the EU Circular Economy Directive, such as specific targets for waste prevention, reuse and specifications on ecodesign. The UK Environmental Audit Committee (2019) "Fixing fashion" [20] focused on EPR as a way to influence product design for disassembly. Finally, the EEA 2019 [21] review of policy shows one of the lowest areas of policy involvement is eco-design both from a market and awareness perspective. Empowering this area of the product life cycle would enable products to have end-of-life planning built in.

4. Testing Active Disassembly in Fashion Design

To investigate the effect of design for disassembly principles on conception and development process, 700 fashion design students will be offered the opportunity to submit a Design for Disassembly Form for each new product they make, disclosing how each item should be dismantled and/or disposed of at the end of its life. The students have been given a full briefing on the context and principles. Follow-up support will be offered to aid student thinking and to discuss how they might make changes to their process, materials or final outcome, or more importantly where they are struggling to produce the desired outcome within the boundaries of design for disassembly. Once all forms are submitted and products made, some students will be invited to follow-up in focus groups where they will be asked to discuss how their overall designs, choice of materials and assembly methods changed. Early results show that alterations are being made to products in order to mitigate landfill waste. This challenge will be taken to e-textiles researchers and development teams, to see if they would make changes to materials, manufacturing methods or overall design. Results from this study will inform discussion around the development of a sustainable waste framework, underpinning policy development for e-textiles waste.

5. Conclusions and Further Study

Figure 1 shows the aspirations of this study where basing policy and legislation for e-textiles on design for disassembly principles will provide product transparency. It will then be clear at 'end of life' how a product can be disposed of and what parts are not recyclable, reusable or biodegradable. Making design for disposal and disassembly central to policy would lead to material and product and process innovation. The overarching policy would be anticipatory, in that e-textiles are a new product without an already established end-of-life story. It would contribute to new dynamic policy principles, filling the current gap in e-textiles waste legislation. If design for disassembly principles are adopted as part of policy design for waste, as a result of this study, it will be evolutionary and directly related to innovation, bringing responsibility to inventors and producers. Once the final tests have been undertaken and analysed, a policy framework will be drafted. This will be tested further through industry focus groups, policy experts and other key stakeholders before being presented as a final policy recommendation.

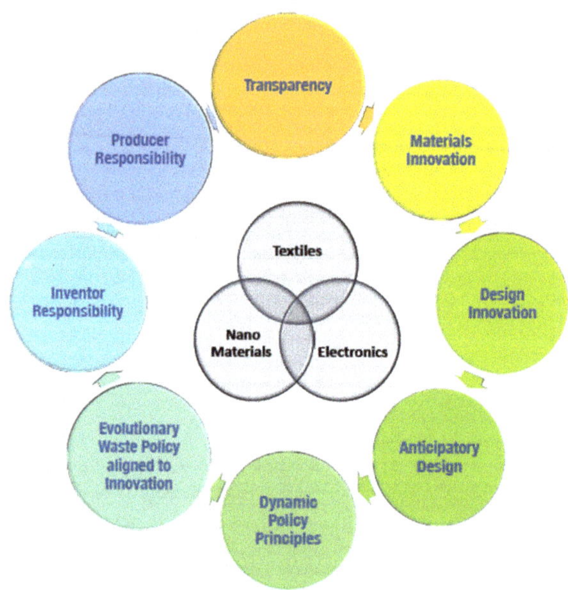

Figure 1. Aims of e-textiles waste design for disassembly waste policy.

Funding: This research received no external funding.

Institutional Review Board Statement: Ethical review and approval are being sought by the UAL ethics committee before undertaking the research that includes participants.

Informed Consent Statement: Informed consent will be obtained from all subjects involved in the study.

Conflicts of Interest: The authors declare no conflict of interest.

References

1. Hayward, J. E-Textiles and Smart Clothing 2020–2030: Technologies, Markets and Players: IDTechEx. 2020. Available online: https://www.idtechex.com/en/research-report/e-textiles-and-smart-clothing-2020-2030-technologies-markets-and-players/735 (accessed on 21 June 2020).
2. Google ATAP. Available online: https://atap.google.com/ (accessed on 1 October 2021).
3. Meinel, C.; Fung, M.L. Clean-IT: Policies to Support Sustainable Digital Technologies. G20 Insights. 2021. Available online: https://www.g20-insights.org/policy_briefs/clean-it-policies-to-support-sustainable-digital-technologies/ (accessed on 19 October 2021).
4. Thomas Faunce and Bartlomiej Kolodziejczyk. Nanowaste: Need for Disposal and Recycling Standards. G20 Insights. December 2020. Available online: https://www.g20-insights.org/policy_briefs/nanowaste-need-disposal-recycling-standards/ (accessed on 19 October 2021).
5. Forti, V.; Baldé, C.P.; Kuehr, R.; Bel, G. The Global E-Waste Monitor 2020: Quantities, Flows and the Circular Economy Potential. United Nations University (UNU)/United Nations Institute for Training and Research (UNITAR)—Co-Hosted SCYCLE Programme, International Telecommunication Union (ITU) & International Solid Waste Association (ISWA), Report, Bonn/Geneva/Rotterdam. 2020. Available online: https://www.itu.int/en/ITUD/Environment/Documents/Toolbox/GEM_2020_def.pdf (accessed on 1 November 2020).
6. Pulse of the Fashion Industry. Common Objective. 2017. Available online: http://www.commonobjective.co/article/pulse-of-the-fashion-industry-2017 (accessed on 2 January 2021).

7. Bentham, J.; Burns, J.H.; Hart, H.L.A. (Eds.) *An Introduction to the Principles of Morals and Legislation (The Collected Works of Jeremy Bentham)*; Clarendon Press: Gloucestershire, UK, 1996.
8. Marciniak, A. "Prevention of evil, production of good": Jeremy Bentham's indirect legislation and its contribution to a new theory of prevention. *Hist. Eur. Ideas* **2017**, *43*, 83–105. [CrossRef]
9. Science for Environment Policy. The Precautionary Priniple: Decision Making under Uncertainty. Future Brief 18. Produced for the European Commission DG Environment by the Science Communication Unit, UWE, Bristol. 2017. Available online: http://ec.europa.eu/science-environment-policy (accessed on 11 July 2021).
10. Kosti, N.; Levi-Faur, D.; Mor, G. Legislation and regulation: Three analytical distinctions. *Theory Pract. Legis.* **2019**, *7*, 169–178. [CrossRef]
11. Eckert, S. Beyond legislation: Reconsidering the locus of power in EU regulatory governance. *Theory Pract. Legis.* **2019**, *7*, 205–225. [CrossRef]
12. Chiodo, J.D.; Ijomah, W.L. Use of active disassembly technology to improve remanufacturing productivity: Automotive application. *Int. J. Comput. Integr. Manuf.* **2014**, *27*, 361–371. [CrossRef]
13. Circular Fashion—A New Textiles Economy: Redesigning Fashion's Future. 2017. Available online: https://www.ellenmacarthurfoundation.org/publications/a-new-textiles-economy-redesigning-fashions-future (accessed on 16 February 2018).
14. WEEE Forum. An Enhanced Definition of EPR and the Role of All Actors. Available online: https://weee-forum.org/wp-content/uploads/2020/11/EPR-and-the-role-of-all-actors_final.pdf (accessed on 13 April 2021).
15. Preston, F. A Global Redesign? Shaping the Circular Economy. Briefing Paper. Chatham House, 2012; p. 20. Available online: https://www.chathamhouse.org/2012/03/global-redesign-shaping-circular-economy (accessed on 2 March 2021).
16. Goldsworthy, K.; Ellams, D. Collaborative Circular Design. Incorporating Life Cycle Thinking into an Interdisciplinary Design Process. *Des. J.* **2019**, *22* (Suppl. 1), 1041–1055. [CrossRef]
17. Bocken, N.M.P.; de Pauw, I.; Bakker, C. Product design and business model strategies for a circular economy. *J. Ind. Prod. Eng.* **2016**, *33*, 308–320. [CrossRef]
18. Moreno, M.; De los Rios, C.; Rowe, Z.; Charnley, F. A Conceptual Framework for Circular Design. *Sustainability* **2016**, *8*, 937. [CrossRef]
19. Expert Network on Textiles Recycling Europe Strategic Agenda ENTeR Project EU 2016–2020. Available online: https://www.interreg-central.eu/Content.Node/Content.Node/Strategic-Agenda.pdf/Stategic-Agenda.pdf (accessed on 21 April 2020).
20. Environmental Audit Committee UK Government. Fixing Fashion: Clothing Consumption and Sustainability—Environmental Audit Committee. House of Commons. 2019. Available online: https://publications.parliament.uk/pa/cm201719/cmselect/cmenvaud/1952/report-files/195207.htm (accessed on 4 February 2020).
21. Manshoven, S.; Christis, M.; Vercalsteren, A.; LaFond, E.; Arnold, M.; Nicolau, M. Textiles and the Environment in a Circular Economy. European Topic Centre Waste and Materials in a Green Economy Eionet Report. ETC/WMGE2019/6, 2019. p. 60. Available online: https://www.eionet.europa.eu/etcs/etc-wmge/products/etc-reports/textiles-and-the-environment-in-a-circular-economy (accessed on 18 May 2021).

engineering proceedings

Proceeding Paper

Green Synthesised Silver Nanocomposite for Thermoregulating E-Textiles [†]

Ashleigh Naysmith [1,*], Naeem S. Mian [2] and Sohel Rana [1]

[1] Department of Fashion and Textiles, University of Huddersfield, Huddersfield HD1 3DH, UK; s.rana@hud.ac.uk
[2] Department of Engineering and Technology, University of Huddersfield, Huddersfield HD1 3DH, UK; n.mian@hud.ac.uk
* Correspondence: ashleigh.naysmith@hud.ac.uk
† Presented at the 3rd International Conference on the Challenges, Opportunities, Innovations and Applications in Electronic Textiles (E-Textiles 2021), Manchester, UK, 3–4 November 2021.

Abstract: Personal thermal management devices provide a behaviourally aligned route to address dependence on energy-intensive heating and cooling systems. E-textiles form an ideal foundation for these devices. In this study, a Joule heating e-textile has been developed using green synthesised silver nanoparticles and polypyrrole, which can easily be dip-coated onto an environmentally benign linen fabric. A Plackett–Burman design was used to optimise the nanoparticle synthesis. Characterisation and electrothermal analysis were carried out to confirm the successful synthesis of silver nanoparticles (40–80 nm, polydispersity index (pdi): 0.25) and an electrical resistance of 28.5 Ω. Joule heating of 66 °C at 6 V applied DC voltage was attained.

Keywords: e-textiles; electronic textile; silver nanoparticles; green synthesis; polypyrrole; personal thermal management (PTM); Joule heating; design of experiment (DOE); Plackett–Burman method; linen

Citation: Naysmith, A.; Mian, N.S.; Rana, S. Green Synthesised Silver Nanocomposite for Thermoregulating E-Textiles. *Eng. Proc.* **2022**, *15*, 15. https://doi.org/10.3390/engproc2022015015

Academic Editors: Steve Beeby, Kai Yang and Russel Torah

Published: 29 April 2022

Publisher's Note: MDPI stays neutral with regard to jurisdictional claims in published maps and institutional affiliations.

1. Introduction

Personal thermal management (PTM) devices are part of an emerging field of technology aimed at addressing the environmental impact of heating and cooling systems, which account for 16% of global energy consumption [1] and 40% of energy-related carbon dioxide emissions [2], whilst also improving the wearer's thermal comfort. E-textiles form an optimal foundation for PTM devices due to their concurrence with thermal comfort adaptation behaviours. Current thermoregulating textile technologies, such as phase change materials [3], provide enhanced thermal comfort, but their functionality is often restricted to when there is an elevation of body temperature. These passive technologies do not allow the user control over their thermal environment. Active thermal comfort strategies aim to overcome the limitations of passive PTM through the use of encapsulated thermistors [4–7], conductive polymers [8–10] and nanomaterials [11–14]. Most current nanomaterial PTM textile research focuses on carbonaceous compounds, which are expensive and lack the conductivity of metallic compounds. Despite the rapid increase in PTM research, to date, there has been no published research demonstrating Joule heating e-textiles based on nanofunctionalised linen fabrics using silver nanoparticle–polypyrrole composites.

Silver nanoparticles (AgNPs) are highly electrically conductive and industrially scalable [15,16]. Polypyrrole is a conjugated polymer that has been selected to enhance the electrical conductivity as well as the adhesion of silver nanoparticles to linen textiles through hydrogen bonding [17]. A linen textile was selected due to its renewability, biodegradability and low energy requirements compared with petrochemical-based textiles, as well as low water consumption compared with cotton [18,19]. Herein, this research demonstrates the development of a green synthesised AgNP–polypyrrole nanocomposite applied to linen through a dip-coating method. Green synthesised nanoparticles commonly suffer from a lack of homogeneity and reliable production. Therefore, a Plackett–Burman design of experiment (DOE) was utilised to identify the optimum parameters for the green synthesis of AgNPs using lime peel extract. The Plackett–Burman method is economical for detecting the main effects, but two-factor interaction is confounded; therefore, it does not distinguish the effect of interactions between variables from the main effects.

2. Materials and Methods

2.1. Materials

Silver nitrate solid ($AgNO_3$) (99% titration, CAS number 7761-88-8), ferric chloride ($FeCl_3$) (CAS number 7705-08-0), pyrrole (>98% assay, CAS number 109-97-7) and sodium carbonate powder (Na_2CO_3) (99.5% assay) were purchased from Sigma-Aldrich. Distilled (DS) water and methanol were used from laboratory stock. All chemicals were received as purchased and used without further purification. Limes were purchased from a local supermarket and washed in tap water prior to use. Valencia natural even-weave linen (approx. 240 gsm) was purchased from Whaleys Bradford.

2.2. Green Synthesis of Silver Nanoparticles

Silver nanoparticles were synthesised according to a modified literature method based on the work of Pugazhenthiran et al. (2021) [20] and Dutta et al. (2020) [21]. Fresh lime peels were cut into small pieces and added to DS water (1:10 w/v) in a borosilicate flask and boiled for 20 min. The extract was then filtered using Fisherbrand filter paper (Grade 601, QL 100, 110 mm diameter). In a typical experiment, silver nitrate solution was added to a conical flask. Lime peel extract (LPE) was added to the solution dropwise using a burette. Once all LPE was added, the pH of the solution was measured and Na_2CO_3 was added to adjust the pH. Parameters were set as per the requirements for the Plackett–Burman protocol. The solution was then centrifuged for 5 min at 3000 RPM to remove any biomass, and the precipitate was discarded. The supernatant was centrifuged (1 h, 6000 RPM, DS water), and the precipitate was rinsed and centrifuged again (20 min, 6000 RPM, methanol). The precipitate was then removed and dried in oven before weighing on an analytical balance.

2.3. Plackett–Burman Optimisation of Synthesis

Minitab software [22] was used to create and analyse a 5 factor, 2 level design with 20 runs and 1 centre point was used.

2.4. In Situ Polymerisation of AgNP Pyrrole on Linen

A linen fabric (even-weave natural, 2.5 cm^2, approx. 0.17 g) was soaked in AgNP solution; then, 0.1 M pyrrole monomer solution was added. The polymerisation initiator ($AgNO_3$ 0.1 M or $FeCl_3$ 0.22 M) was next added to the solution and covered to protect it from light. The polymerisation reaction was carried out for up to 96 h.

2.5. Characterisation

AgNP formation was confirmed via UV–Vis spectroscopy (UV/Vis/NIR Spectrophotometer, Jasco V-700 series). AgNP size and polydispersity were analysed using dynamic light scattering (DLS) (Malvern Zetasizer Nano ZS). Elemental analysis was conducted using an energy dispersive X-ray fluorescence spectrometer (EDX-XRF) (Shimadzu EDX-7000 XRF). Resistance was measured using an LCR bridge (Rohde and Schwarz HM8118). Joule heating performance was tested using an FLIR thermal camera whilst the linen fabric was connected to a DC voltage supply via crocodile clips.

3. Results and Discussion

3.1. Silver Nanoparticle Morphological Analyses

AgNP synthesis was confirmed through spectral UV–Vis analysis by exhibiting the characteristic surface plasmon absorption maxima between 391 nm and 453 nm [23]. AgNPs undergo surface plasmon resonance, causing an absorption of light at higher wavelengths than non-plasmonic particles of the same size [24]. A secondary peak, or very broad peak, can be seen where AgNPs have agglomerated. As expected, many of the parameters did not form AgNPs, with broad peaks indicating agglomeration and peaks with absorption maxima at long wavelengths indicating oxidation of the silver.

The results from the 21 DOE runs were analysed through Minitab, and the response optimiser function was employed to produce a solution for optimal AgNP synthesis using LPE. AgNPs were synthesised using the Minitab solution parameters (pH $\pm$ 0.7). The resultant silver nanoparticles produced a UV–Vis peak at 420 nm with a diameter of 117 nm and a polydispersity index of 0.243 (Figure 1). The absorption maxima peak and polydispersity index were close to the predicted solution, and subsequent experiments were repeatable. However, the particle size (diameter/nm) obtained from the zeta sizer was larger than predicted. This could be due to the effect of pH on the size and addition of sodium carbonate not providing the required degree of precision, or the agglomeration of the particles during analysis. A repeat experiment would benefit from using a wider range of Na_2CO_3 solutions with different molarities to finetune the pH adjustment. The pH of the reaction directly impacts on the size and stability of AgNPs, with higher acidity syntheses tending to produce more agglomerated AgNPs [25]. The polydispersity index of 0.243 indicates a monodisperse collection of AgNPs [25].

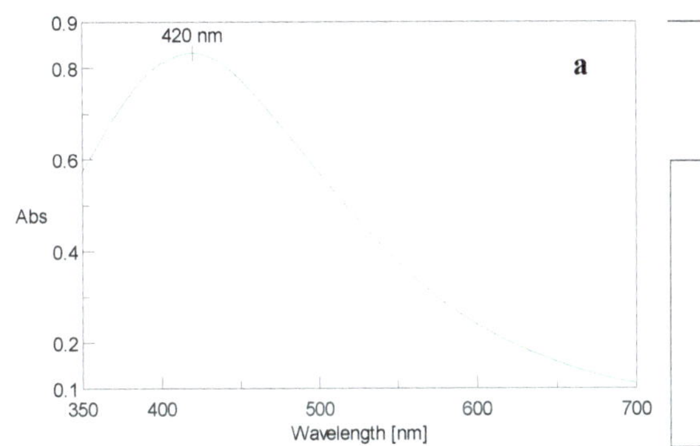
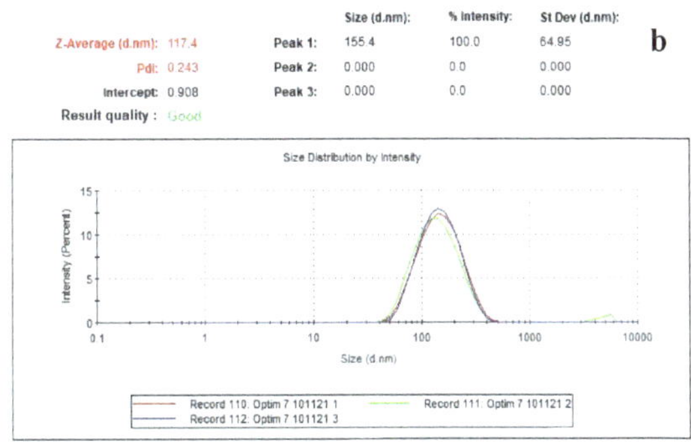

Figure 1. (**a**) UV–Vis spectra for optimised nanoparticles, (**b**) DLS analysis for optimised silver nanoparticles.

3.2. Thermoelectric Performance of Polypyrrole Silver Nanoparticle Linen

The electrical resistance of the developed polypyrrole–AgNP linen (PAL) reached a very low value of 28.5 Ω (at 1 mm thickness, this is equivalent to 285 Ω/sq.) using optimised AgNPs with 0.1 M pyrrole and 0.1 M $AgNO_3$ (PAL-$AgNO_3$-0). This is comparable to the values reported in state-of-the-art thermoregulatory nanomaterial-based e-textiles which range from 0.16 Ω/sq [12] to 1.16 $\times$ 10^4 Ω/sq [26]. Joule heating analysis revealed that PAL-$AgNO_3$-0 reached a temperature of 66 °C with 6 V applied DC voltage (Figure 2). The electrical resistance of PAL using the optimised AgNPs with 0.1 M pyrrole and 0.22 M $FeCl_3$ (PAL-$FeCl_3$-0) was 395.36 Ω, and the Joule heating performance was 28.7 °C with 6 V applied DC voltage. Polypyrrole-coated linen fabrics without AgNPs exhibited an electrical resistance of 421 Ω and reached a temperature of 24.5 °C at 6 V DC input.

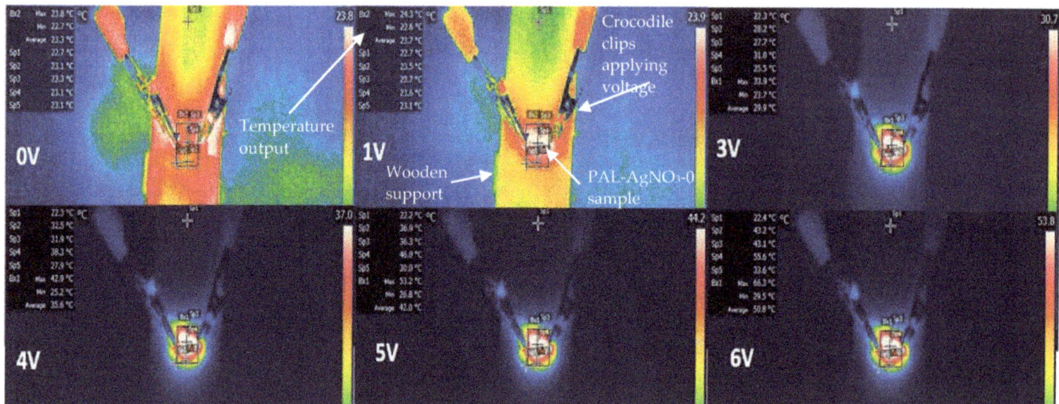

Figure 2. Thermal camera images for polypyrrole–silver nanoparticle linen at various applied voltages.

3.3. EDX-XRF Analysis

Elemental analysis for PAL-AgNO$_3$-0 revealed that the inorganic compounds present on the surface of the fabrics were traces of calcium (0.004%) and copper (0.001%), and no silver was detected. However, the electrothermal performance of PAL-AgNO$_3$-0 significantly exceeded that of polypyrrole-only linen. PAL-FeCl$_3$-0 exhibited the presence of 0.9% silver in the EDX spectra and several trace elements. Further research is under way to explore this finding.

4. Conclusions

In this study, AgNPs have been successfully synthesised using lime peel extract as the reducing agent, providing a simple and low-cost route. The synthesis has been optimised using a Plackett–Burman experimental design to identify key parameters enabling the synthesis to be tuned to reliably obtain the desired size of AgNPs. A nanocomposite was subsequently developed using the green synthesised AgNPs and polypyrrole polymerised in situ on linen fabrics to obtain a linen-based e-textile that demonstrates a low electrical resistance of 28.5 Ω and a Joule heating performance of 66 °C at 6 V applied DC voltage. Further optimisation studies are under way; future work will analyse the results of this study and investigate the temperature-dependent electrical resistance, mechanical properties, and morphology of the developed e-textiles.

Author Contributions: A.N. conceived the idea, conducted the experiments, and wrote the paper. S.R. and N.S.M. reviewed and edited the manuscript. S.R. and N.S.M. supervised and administered the project. All authors have read and agreed to the published version of the manuscript.

Funding: This research was funded by the UK Engineering and Physical Sciences Research Council (EPSRC) grant EP/T51813X/1 for the University of Huddersfield Doctoral Training Programme.

Institutional Review Board Statement: Not applicable.

Informed Consent Statement: Not applicable.

Data Availability Statement: Not applicable.

Acknowledgments: The authors would like to thank Kay Burrows, Timothy Smith, Andrew Hebden, and Leslie Johnson for their support and assistance throughout the research; Ali Iqbal for his assistance with thermal analyses; Ibrahim George and James Rooney for their assistance with EDX; and Hayley Markham for her assistance with zeta analysis. Steven Gallacher, Martin Webster, and David Bray carried out the electrical resistance testing.

Conflicts of Interest: The authors declare no conflict of interest.

References

1. Ürge-Vorsatz, D.; Cabeza, L.F.; Serrano, S.; Barreneche, C.; Petrichenko, K. Heating and cooling energy trends and drivers in buildings. *Renew. Sustain. Energy Rev.* **2015**, *41*, 85–98. [CrossRef]
2. Smith, A. *A Net Zero Climate-Resilient Future: Science, Technology and the Solutions for Change*; The Royal Society: London, UK, 2021.
3. Kou, Y.; Sun, K.; Luo, J.; Zhou, F.; Huang, H.; Wu, Z.-S.; Shi, Q. An intrinsically flexible phase change film for wearable thermal managements. *Energy Storage Mater.* **2021**, *34*, 508–514. [CrossRef]
4. Hughes-Riley, T.; Jobling, P.; Dias, T.; Faulkner, S.H. An investigation of temperature-sensing textiles for temperature monitoring during sub-maximal cycling trials. *Text. Res. J.* **2020**, *91*, 624–645. [CrossRef]
5. Hughes-Riley, T.; Dias, T.; Cork, C. A Historical Review of the Development of Electronic Textiles. *Fibers* **2018**, *6*, 34. [CrossRef]
6. Morris, N.B.; Jay, O.; Flouris, A.D.; Casanueva, A.; Gao, C.; Foster, J.; Havenith, G.; Nybo, L. Sustainable solutions to mitigate occupational heat strain—An umbrella review of physiological effects and global health perspectives. *Environ. Health* **2020**, *19*, 95. [CrossRef]
7. Lugoda, P.; Hughes-Riley, T.; Oliveira, C.; Morris, R.; Dias, T. Developing Novel Temperature Sensing Garments for Health Monitoring Applications. *Fibers* **2018**, *6*, 46. [CrossRef]
8. Lee, J.-W.; Han, D.-C.; Shin, H.-J.; Yeom, S.-H.; Ju, B.-K.; Lee, W. PEDOT:PSS-Based Temperature-Detection Thread for Wearable Devices. *Sensors* **2018**, *18*, 2996. [CrossRef]
9. Wang, B.; Cheng, H.; Zhu, J.; Yuan, Y.; Wang, C. A flexible and stretchable polypyrrole/knitted cotton for electrothermal heater. *Org. Electron.* **2020**, *85*, 105819. [CrossRef]
10. Zhang, L.; Baima, M.; Andrew, T.L. Transforming Commercial Textiles and Threads into Sewable and Weavable Electric Heaters. *ACS Appl. Mater. Interfaces* **2017**, *9*, 32299–32307. [CrossRef]
11. Ahmed, A.; Jalil, M.A.; Hossain, M.M.; Moniruzzaman, M.; Adak, B.; Islam, M.T.; Parvez, M.S.; Mukhopadhyay, S. A PEDOT:PSS and graphene-clad smart textile-based wearable electronic Joule heater with high thermal stability. *J. Mater. Chem. C* **2020**, *8*, 16204–16215. [CrossRef]
12. Guo, Z.; Sun, C.; Zhao, J.; Cai, Z.; Ge, F. Low-Voltage Electrical Heater Based on One-Step Fabrication of Conductive Cu Nanowire Networks for Application in Wearable Devices. *Adv. Mater. Interfaces* **2020**, *8*, 2001695. [CrossRef]
13. Ke, F.; Song, F.; Zhang, H.; Xu, J.; Wang, H.; Chen, Y. Layer-by-layer assembly for all-graphene coated conductive fibers toward superior temperature sensitivity and humidity independence. *Compos. Part B: Eng.* **2020**, *200*, 108253. [CrossRef]
14. Kim, H.; Lee, S. Effect of Relative Humidity Condition on Electrical Heating Textile Coated with Graphene-based on Cotton Fabric. *Fibers Polym.* **2021**, *22*, 276–284. [CrossRef]
15. Babaahmadi, V.; Montazer, M.; Gao, W. Low temperature welding of graphene on PET with silver nanoparticles producing higher durable electro-conductive fabric. *Carbon* **2017**, *118*, 443–451. [CrossRef]
16. Hazarika, A.; Deka, B.K.; Kim, D.Y.; Park, Y.-B.; Park, H.W. Smart gating of the flexible Ag@CoxMo1-xP and rGO-loaded composite based personal thermal management device inspired by the neuroanatomic circuitry of endotherms. *Chem. Eng. J.* **2020**, *421*, 127746. [CrossRef]
17. Zhao, H.; Hou, L.; Lu, Y. Electromagnetic interference shielding of layered linen fabric/polypyrrole/nickel (LF/PPy/Ni) composites. *Mater. Des.* **2016**, *95*, 97–106. [CrossRef]
18. Shen, L.; Patel, M.K. Life Cycle Assessment of Polysaccharide Materials: A Review. *J. Polym. Environ.* **2008**, *16*, 154. [CrossRef]
19. Werf, H.M.G.v.d.; Turunen, L. The environmental impacts of the production of hemp and flax textile yarn. *Ind. Crops Prod.* **2008**, *27*, 1–10. [CrossRef]
20. Pugazhenthiran, N.; Murugesan, S.; Muneeswaran, T.; Suresh, S.; Kandasamy, M.; Valdés, H.; Selvaraj, M.; Dennyson Savariraj, A.; Mangalaraja, R.V. Biocidal activity of citrus limetta peel extract mediated green synthesized silver quantum dots against MCF-7 cancer cells and pathogenic bacteria. *J. Environ. Chem. Eng.* **2021**, *9*, 105089. [CrossRef]
21. Dutta, T.; Ghosh, N.N.; Das, M.; Adhikary, R.; Mandal, V.; Chattopadhyay, A.P. Green synthesis of antibacterial and antifungal silver nanoparticles using Citrus limetta peel extract: Experimental and theoretical studies. *J. Environ. Chem. Eng.* **2020**, *8*, 104019. [CrossRef]
22. Minitab. Available online: https://app.minitab.com (accessed on 16 December 2021).
23. Amirjani, A.; Firouzi, F.; Haghshenas, D.F. Predicting the Size of Silver Nanoparticles from Their Optical Properties. *Plasmonics* **2020**, *15*, 1077–1082. [CrossRef]
24. Paramelle, D.; Sadovoy, A.; Gorelik, S.; Free, P.; Hobley, J.; Fernig, D.G. A rapid method to estimate the concentration of citrate capped silver nanoparticles from UV-visible light spectra. *Analyst* **2014**, *139*, 4855–4861. [CrossRef] [PubMed]
25. Tarannum, N.; Divya, D.; Gautam, Y.K. Facile green synthesis and applications of silver nanoparticles: A state-of-the-art review. *RSC Adv.* **2019**, *9*, 34926–34948. [CrossRef]
26. Rosace, G.; Trovato, V.; Colleoni, C.; Caldara, M.; Re, V.; Brucale, M.; Piperopoulos, E.; Mastronardo, E.; Milone, C.; Luca, G.D.; et al. Structural and morphological characterizations of MWCNTs hybrid coating onto cotton fabric as potential humidity and temperature wearable sensor. *Sens. Actuators B Chem.* **2017**, *252*, 428–439. [CrossRef]

Proceeding Paper

An All Dispenser Printed Electrode Structure on Textile for Wearable Healthcare †

Meijing Liu *, Zeeshan Ahmed, Neil Grabham, Stephen Beeby, John Tudor and Kai Yang

Smart Electronic Materials and Systems Research Group, School of Electronics and Computer Science, University of Southampton, Southampton SO17 1BJ, UK; zeeshan1@live.co.uk (Z.A.); njg@ecs.soton.ac.uk (N.G.); spb@ecs.soton.ac.uk (S.B.); mjt@ecs.soton.ac.uk (J.T.); ky2e09@soton.ac.uk (K.Y.)
* Correspondence: ml5y17@soton.ac.uk
† Presented at the 3rd International Conference on the Challenges, Opportunities, Innovations and Applications in Electronic Textiles (E-Textiles 2021), Manchester, UK, 3–4 November 2021.

Abstract: This paper presents a dispenser printed electrode array on polyester/cotton fabric. The fabrication details needed to achieve the array, including the materials and printer set-up, are reported. The array consists of ten electrode elements for functional electrical stimulation (FES), including nine active electrodes and one common return electrode. The minimum gap between conductive tracks of 1 mm required for the design was achieved. The fabrication method can be used to tailor the electrode array to fit a wide variety of healthcare applications and an individual's requirements for personalized healthcare.

Keywords: dispenser printing; wearable electrode; E-textile; fabrication process

Citation: Liu, M.; Ahmed, Z.; Grabham, N.; Beeby, S.; Tudor, J.; Yang, K. An All Dispenser Printed Electrode Structure on Textile for Wearable Healthcare. *Eng. Proc.* **2022**, *15*, 16. https://doi.org/10.3390/engproc2022015016

Academic Editors: Steve Beeby and Russel Torah

Published: 29 April 2022

Publisher's Note: MDPI stays neutral with regard to jurisdictional claims in published maps and institutional affiliations.

1. Introduction

Screen-printed dry electrodes for electrotherapy applications have been demonstrated in previous work [1]. However, dispenser printing offers the advantage of lower cost for custom designs compared to screen printing, since no physical screen is required for patterning. It is a direct-write printing technique where the material structures are built via a syringe and nozzle connected to a pneumatic pressure controller [2]. The printing pattern is automatically controlled by the computer software according to the required design, which makes bespoke designs feasible.

Electrodes are key components of many healthcare devices for diagnostics/monitoring (e.g., ECG, EEG, EMG) and therapeutics (e.g., electrical muscle stimulation for assisted living and rehabilitation). Traditional hydrogel electrodes are not suitable for long-term use because their performance deteriorates over time due to moisture evaporation and contamination build-up resulting from their stickiness. A dry electrode made from Fabink E-0002 paste was used in textile printing in previous work for healthcare applications [3]. However, the pot-life of Fabink E-0002 is only around 20 min, which is often not sufficient for the fabrication process. The short pot-life of the ink occurs since the viscosity of the paste increases due to crosslinking of the electrode components after mixing. This also creates a challenge during dispenser printing because the syringe pressure needs to be adjusted as the ink's viscosity increases during printing. This paper introduces and describes the application of a new dry electrode paste called Fabink E-0003. The pot-life of Fabink E-0003 is around three hours.

This work presents the fabrication process and parameters of an electrode array on a textile by dispenser printing. The strategy for interface printing and printing characterization is discussed.

2. Methods and Materials

2.1. Dispenser Printer

Dispenser printing uses air pressure to dispense inks out of a nozzle onto a substrate. The Fisnar F7300NV 3 Axis dispenser printer used in this study is shown in Figure 1a. The nozzle can move in the X and Z directions relative to the substrate, and Y direction movement can be achieved by the substrate moving relative to the nozzle. The movement of both the nozzle and the substrate is controlled by a computer system. The movement instructions for the printing of each layer are controlled by the following parameters: printing speed, line gap and nozzle height above the substrate. Line gap is the distance between two parallel lines of a meander printing trace as shown in Figure 1b. The nozzle can be changed manually to the appropriate size determined by the required print resolution; available nozzle diameters range from 0.15 mm to 1.60 mm. The air pressure is set by the pressure controller depending on the viscosity of the material to be dispensed with a higher viscosity requiring a higher pressure.

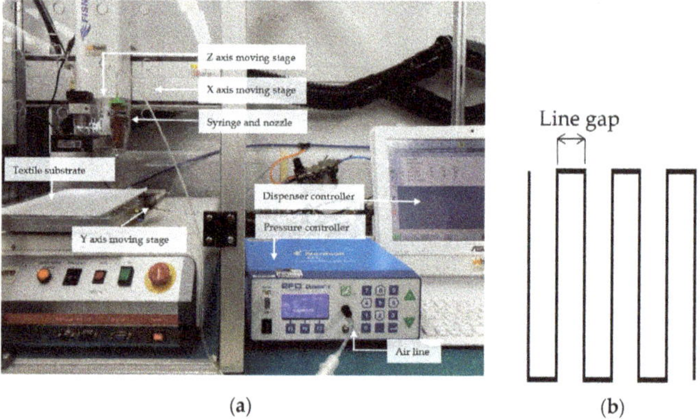

(a)

(b)

Figure 1. (**a**) The Fisnar F7300NV 3 Axis dispenser printer. (**b**) An example of meander printing to achieve a square. Line gap is the distance between two parallel lines.

The electrode array for functional electrical stimulation (FES) consists of ten electrode elements comprising nine active electrodes and one common return electrode. As in previous work [4], the sample consists of four functional layers, as shown in Figure 2a–d. Figure 2e shows the final top view of the sample after depositing four layers. The minimum distance between two silver tracks is designed to be 1 mm, which is also the width of the printed conductive tracks.

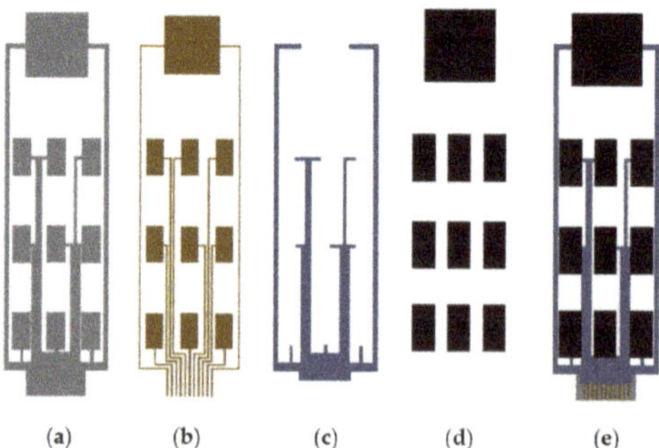

(a)   (b)   (c)   (d)   (e)

Figure 2. The design of each layer: (**a**) interface layer; (**b**) conductive silver layer; (**c**) encapsulation layer; (**d**) carbon layer; (**e**) the top view of the sample with all four layers completed.

2.2. Materials

The woven textile substrate is polyester/cotton A1656 white supplied by Whaleys Bradford Ltd. (Bradford, UK). The three functional pastes used in this study are listed in Table 1. They were supplied by Smart Fabric Inks Ltd. (Southampton, UK). The interface, silver and encapsulation material are the same as those used in previous work [1]. However, the electrode paste changed to Fabink E-0003.

Table 1. Properties and curing conditions of pastes.

Pastes	Functionality	Curing Conditions
Fabink UV-IF-1004	Interface and encapsulation to create a smooth surface and electrical insulation	UV light, 60 s
Fabink TC-C4007	Silver ink for printing flexible conductive layer	100 °C, 30 min
Fabink E-0003	Carbon paste for printing electrodes	120 °C, 30 min

3. Fabrication Processes

The dispenser printed electrode array consists of four functional layers as shown in Figure 2. The set-up of the printer for each layer is optimized, and the optimum parameters are listed in Table 2.

Table 2. Dispenser printer parameters for each layer.

Functional Layer	Layer Number	Nozzle Size (mm)	Pressure (k Pa)	Line Gap (mm)	Printing Speed (mm/s)	Height (mm)
Interface	1	1.6	15	1	35	0.05
	2	1.6	15	1	35	0.05
	3	0.41	30	0.2	35	0.05
	4	0.41	30	0.2	35	0.5
	5	0.41	60	0.2	35	1
Silver	6	0.41	20	0.2	15	0.2
Encapsulation	7	1.6	15	1	35	0.1
Carbon electrode	8	1.6	450–550	1	2	1.3

(1) Interface

The function of the interface is to fill any gaps in the textile (e.g., between yarns) and create a smooth and flat surface before printing the subsequent layers. As listed in Table 2, layers 1 and 2 are printed to fill the gaps between textile yarns and create a continuous layer on the textile. Due to the liquid absorption property of the textile, the paste spreads on the bare textile before it is cured. A longer processing time before curing leads to more bleeding. The line gap of the first two prints is set at 1 mm using a nozzle size of 1.6 mm. This saves printing time compared to when a 0.2 mm line gap is used with a smaller nozzle size (0.41 mm) in later layers. Therefore, the paste bleeding is reduced as the printing time is shorter.

After the first two interface layers are printed, the ink forms an uneven surface on the textile. Layer 3 is printed to smooth the surface by filling in the areas of lower height in the uneven surface. A 0.05 mm height is set between the nozzle and the surface peak. Layers 4 and 5 are printed to achieve the additional filling of the lower height areas, resulting in a flat surface. A line gap of 0.2 mm was used for layers 3 to 5. Dividing one wide printing track into five narrower tracks can create a less-curved surface, as illustrated in Figure 3.

Figure 3. The effect of line gap on printing result. (**a**) The cross-section of a printed track achieved by one wide print. (**b**) The cross-section of a printed track achieved by five narrower prints.

The interface layer was formed using Fabink IF-UV-1004. UV curing was applied after each print by exposing the sample to a 400 W mercury (Hg) bulb in a UV cabinet supplied by UV Light Technology Ltd. (Birmingham, UK). The UV curing time was 60 s.

(2)   Silver

The silver layer is printed to form the conductive tracks for interconnections and the conductive pads for the carbon electrodes. The line gap is set as 0.2 mm, which means that a silver track of 1 mm width is achieved by five parallel line prints. Dividing one line into five prints can also reduce the frequency of open circuits leading to improved sample yield and quality. The conductive layer was formed using Fabink TC-C4007. The paste was cured in a box oven at 100 °C for 30 min.

(3)   Encapsulation

The encapsulation layer is printed to protect the conductive tracks and provide electrical insulation. The encapsulation layer was formed using Fabink IF-UV-1004 and cured as for the interface.

(4)   Carbon electrode array

Carbon electrodes, which form the contact layer to the skin, are printed on top of the conductive pads. The electrode paste was cured in a box oven at 120 °C for 30 min resulting in an electrode layer with a thickness of 1.3 mm.

4. Results and Discussion

The samples after printing each of the functional layers are shown in Figure 4. Figure 5a is an SEM image showing the cross-section of the conductive layer printed on the textile.

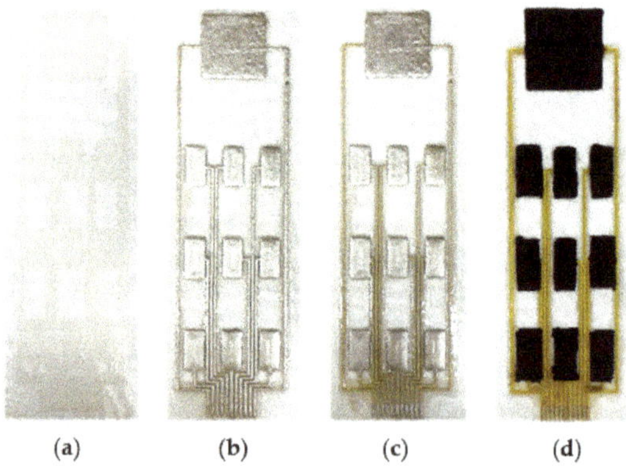

Figure 4. Samples after each printing stage: (**a**) interface layer; (**b**) silver layer; (**c**) encapsulation layers; (**d**) electrode layer.

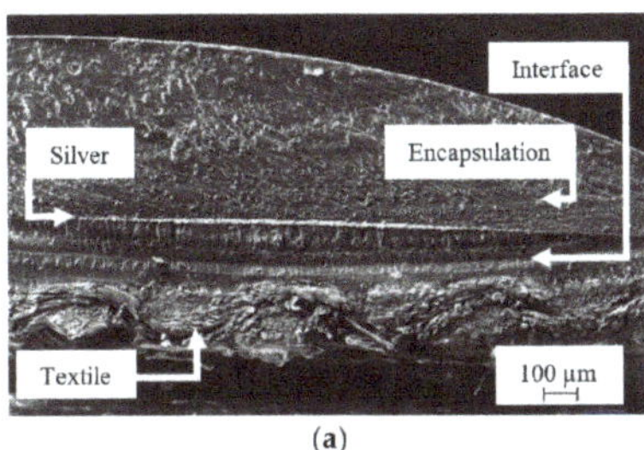

(a)　(b)

Figure 5. (**a**) SEM images showing the cross section of one silver line sandwiched between the interface and encapsulation. (**b**) The SEM image of the textile cross section.

Compared with the bare uncoated textile, which is shown in Figure 5b, the surface roughness is decreased by the interface layer. The curved surface caused by surface tension can be observed in the surface of the encapsulation layer in Figure 5a. In contrast, the interface layer and silver layer are flat with a less-curved surface.

5. Conclusions

This paper has achieved an all dispenser printed electrode array structure on textile. Dispenser printing is a direct write process, which allows the design of the array to be adjusted to achieve bespoke one-off designs tailored to suit an individual's needs. The minimum gap between conductive tracks of 1 mm required in the design was achieved. The process details and materials used are presented. The array consists of ten electrode elements for functional electrical stimulation (FES), including nine active electrodes and one common return electrode. Future work will test the electrode array for muscle stimulation.

Author Contributions: Conceptualization, N.G., J.T. and K.Y.; methodology, Z.A., J.T. and K.Y.; software, Z.A. and M.L.; validation, M.L.; writing—original draft preparation, M.L.; writing review and editing, J.T. and K.Y.; visualization, M.L.; supervision, S.B. and K.Y.; project administration, J.T. and K.Y.; funding acquisition, J.T. and K.Y. All authors have read and agreed to the published version of the manuscript.

Funding: This research was funded by EPSRC under grant number EP/S001654/1.

Institutional Review Board Statement: Not applicable.

Informed Consent Statement: Not applicable.

Data Availability Statement: The data presented in this study are openly available in FigShare at https://doi.org/10.6084/m9.figshare.19552288.v1.

Conflicts of Interest: S.B., J.T. and K.Y are Directors of Smart Fabric Inks Limited.

References

1. Liu, M.; Glanc-Gostkiewicz, M.; Beeby, S.; Yang, K. Fully Printed Wearable Electrode Textile for Electrotherapy Application. *Proceedings* **2021**, *68*, 12. [CrossRef]
2. Greig, T.; Torah, R.; Yang, K. Investigation of Nozzle Height Control to Improve Dispenser Printing of E-Textiles. *Proceedings* **2021**, *68*, 6. [CrossRef]
3. Liu, M.; Ward, T.; Young, D.; Matos, H.; Wei, Y.; Adams, J.; Yang, K. Electronic textiles based wearable electrotherapy for pain relief. *Sens. Actuators A Phys.* **2020**, *303*, 111701. [CrossRef]
4. Yang, K.; Meadmore, K.; Freeman, C.; Grabham, N.; Hughes, A.; Wei, Y.; Torah, R.; Glanc-Gostkiewicz, M.; Beeby, S.; Tudor, J. Development of user-friendly wearable electronic textiles for healthcare applications. *Sensors* **2018**, *18*, 2410. [CrossRef] [PubMed]

engineering proceedings

Proceeding Paper

Investigating the Mechanical Failures at the Bonded Joints of Screen-Printed E-Textile Circuits [†]

Abiodun Komolafe * and Russel Torah

Centre of Flexible Electronics and E-Textiles, University of Southampton, Southampton SO17 1BJ, UK; rnt@ecs.soton.ac.uk

* Correspondence: a.o.komolafe@soton.ac.uk

† Presented at the 3rd International Conference on the Challenges, Opportunities, Innovations and Applications in Electronic Textiles (E-Textiles 2021), Manchester, UK, 3–4 November 2021.

Abstract: It is often necessary to connect e-textile devices with power supplies and other peripherals using electrical wires. This connection is usually achieved with the use of wires that are consequently bonded to the e-textile circuit using conductive epoxies or solders. This paper reports the mechanical failures that arise from this bonded joint during bending by considering the connection of textile-based Litz wires to screen-printed silver conductors using a combination of conductive epoxies and tapes as bonding adhesives. Cyclic bending results of the conductors around a 5 mm bending diameter rod show that conductors with bonded joints degrade after 3500 cycles with the formation of cracks and fractures around the bonded joints. Conductors without bonded joints achieve more than 10,000 bending cycles without the formation of cracks in the conductors.

Keywords: e-textiles; bonded joints; cyclic bending; mechanical failures; connectors; screen printing

Citation: Komolafe, A.; Torah, R. Investigating the Mechanical Failures at the Bonded Joints of Screen-Printed E-Textile Circuits. *Eng. Proc.* **2022**, *15*, 17. https://doi.org/10.3390/engproc2022015017

Academic Editors: Steve Beeby and Kai Yang

Published: 12 May 2022

Publisher's Note: MDPI stays neutral with regard to jurisdictional claims in published maps and institutional affiliations.

1. Introduction

E-textiles combine electronic and textile materials to achieve electronic solutions that are more integrated into human life. Such solutions include health monitoring and rehabilitation [1], aesthetic augmentation [2], and sustainable energy solutions for low power wearable devices [3]. However, the transition of e-textiles from research prototypes to commercial devices significantly hinges on the reliability and robustness of the e-textile in real consumer environments [4]. One such reliability concern centres on achieving durable wired connections between e-textile circuits and their power supplies and/or external peripheral devices. Most of the reliability research on e-textiles has focused on preventing fatigue failures along the embedded conductors under practical stresses. Different methods for improving the durability of conductors that have been reported include the optimization of conductor design with serpentine structures [5], locating the conductors on the neutral axis of the printed e-textile [6], and using mechanically resilient functional materials to enhance the lifetime of the e-textile [7,8]. However, the first point of mechanical failure typically occurs at the bonded joints where electronic components are bonded to the embedded conductors [9] or where the wires connect the e-textiles to external devices and power supplies [6]. For a typical printed e-textile construction, shown in Figure 1, wired connections are unavoidable, hence it is imperative to find solutions that mitigate these failures through robust bonding methods.

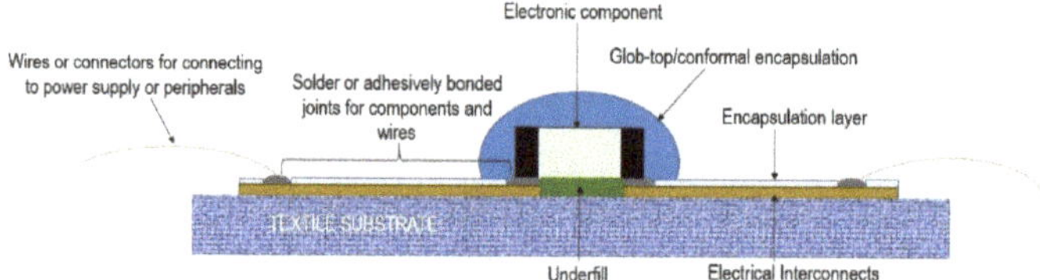

Figure 1. A typical construction of printed e-textiles.

This paper evaluates three different methods for joining textile-based Litz wires to screen printed silver conductors on a polyester fabric, as shown in Figure 2. The bonded joints shown in Figure 2 consist of rigid, flexible and intermediate joints implemented by using conductive films and adhesives. The durability of these joints is assessed through the real-time acquisition and logging of the resistance change of the printed conductor under repeated cyclic bending. This approach simulates the practical bending stresses associated with the use of textiles and examines the behaviour of the bonded joints under mechanical stress. This study is useful for understanding the challenges associated with interfacing e-textile circuits and devices with required peripherals.

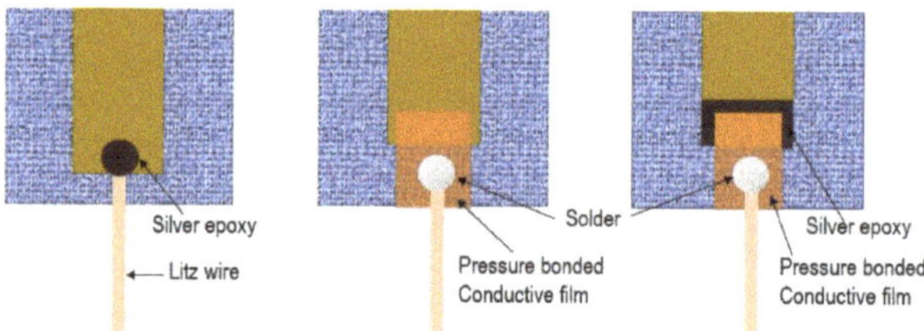

Figure 2. Bonding methods for wired connections on screen printed e-textiles showing (from left to right) rigid, flexible, and intermediate bonded joints.

2. Materials and Methods

The screen-printed e-textile shown in Figure 3 uses a dumbbell-shaped interface and conductor layer pattern and was fabricated on a DEK 248 semi-automatic screen printer based on the printing process described in [10]. The non-planar surface of textiles often necessitates the printing of an interface layer to smooth and planarize the textile before any functional layers can be printed. Consequently, in this work a custom-made polyester fabric, IsacordPoly60, with an average thickness of 207 μm was used as the textile substrate [11]. The textile was selected to limit the printed thickness of the interface layer to 50 μm with a surface roughness of <2 μm, which currently represents the state of the art. The interface and conductor layers were realized using screen printable UV cured polyurethane ink, Fabink UV-IF1004, and thermally cured silver paste, Fabinks TC-4007, from Smart Fabric Inks Ltd. These inks were chosen for their printability, flexibility, and strong adhesion to textiles [12]. The achieved thickness for the silver conductor was 5 μm.

Figure 3. Screen-printed dumb-bell interface (**left**) and silver conductor (**right**) patterns on a bespoke woven Isacord Polyester 60 thread count (IsacordPoly60) fabric.

Silver loaded epoxy, conductive copper film, and low-temperature solders were used to attach textile-based Litz wire to the printed conductor based on the three bonding methods shown in Figure 2. The silver epoxy was cured for 10 min at 120 °C while the copper film was pressure-bonded onto the conductor after the Litz wires were soldered. Since the silver-loaded epoxy becomes stiff and rigid when fully cured, the conductive copper film was chosen to increase the flexibility of the bonded joints and minimize the stiffness gradient between the bonding material and the printed conductor.

Test Setup for Monitoring Bonded Joint during Cyclic Bending

Three samples of each of the bonding methods were subjected to 90° cyclic bending around a 5 mm bending radius under a bending tensile load of 1.5 N as described in [13]. The cyclic bending is driven by a stepper motor. Real-time monitoring and assessment of the performance of these bonded joints under cyclic bending stress was achieved by continuously measuring and recording the electrical resistance of the screen-printed conductor using a LabView controlled Keithley multimeter, as shown in Figure 4. The durability of the samples containing any of the bonding methods was compared with that of samples without any bonding material.

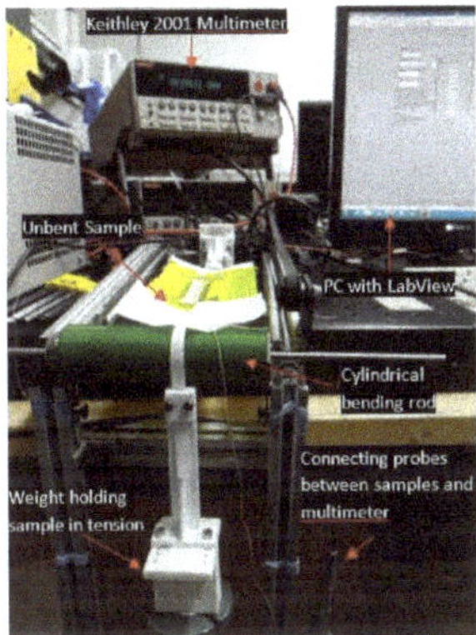

Figure 4. Bending test setup for real-time monitoring of bonded joints under cyclic bending.

3. Results and Discussion

The result showed that the presence of the bonded joints generally induced more stress on the silver conductors during bending and generated the fractures shown in Figure 5. Whilst samples without the bonded joints survived more than 10,000 cycles before crack formation, samples with bonded joints only survived 3500 cycles, after which the electrical resistance of the samples could not be measured by the Keithley multimeter. These fractures introduced poor contact between the contact pad of the conductor and the bonding materials. The poor contact triggered high contact resistance between the wire and the silver conductor and introduced noise into the electrical resistance measurement of the printed conductor, as shown in Figure 6. Figure 6a shows the actual change in the normalised electrical resistance (i.e., the ratio of the electrical resistance in bending to the measured resistance before bending) of the printed conductor over 3000 bending cycles before any noticeable failure at the bonded joint of the conductive epoxy was detected. As soon as the bonded joint began to fail, an upsurge of almost 1000 was initially noticed in the normalised resistance. This eventually deteriorated with an increase in the normalized resistance of up to 7000 within 1000 bending cycles, as shown in Figure 6b.

Figure 5. Crack formation on samples with bonded joints conductive epoxy (**left**), pressure bonded conductive film (**middle**) and epoxy + conductive film (**right**).

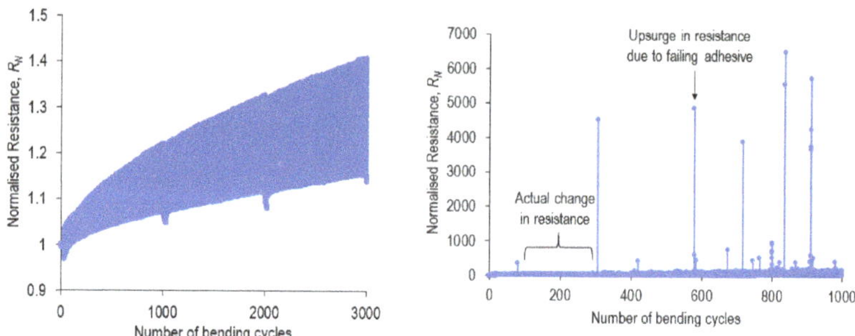

Figure 6. Normalised resistance changes of samples using conductive epoxy before sudden spike resistance measurement (**left**) and during the failure of the bonded joint (**right**).

Figure 7 compares the performance of the flexible bonded joints created from electrically conductive film, and the intermediate joints formed by combining conductive epoxy with the conductive film. The results indicate that the flexible joints quickly degraded under cyclic stress as shown in Figure 7a and introduced sudden peaks and fluctuations to the measurement within the first 1000 bending cycles due to intermittent contact. The magnitude of this generated noise reduced with the intermediate joints, which used the conductive epoxy to increase the bonding strength of the conductive film as shown in Figure 7b. This result was expected since the epoxy-based bonded joint showed more resilience under bending. In all the tested samples, the failure point always occurred at the interface between the bonding material and the printed conductor, as shown in Figure 5. The crack length propagated through the printed conductor when the sample was repeatedly loaded with bending stress.

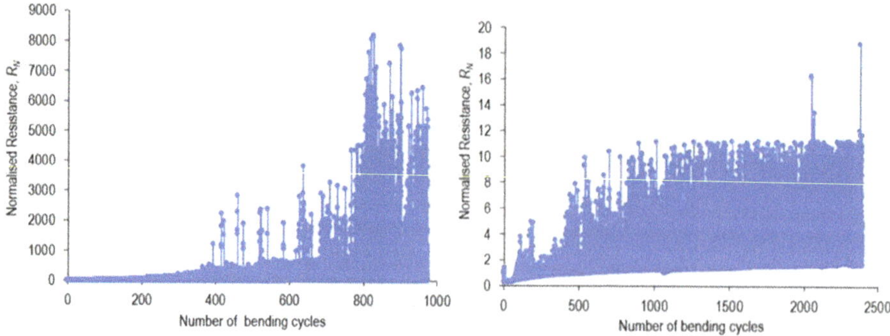

Figure 7. Normalised resistance changes of samples using pressure-bonded conductive film (**left**) and conductive epoxy + conductive film (**right**).

4. Conclusions

Bonded joints are currently unavoidable for e-textile design because they are useful for interconnecting devices on the textile and for wiring/connecting e-textiles to power supplies and peripherals. This paper reports the durability of three different bonded joints of varied degrees of flexibility. The results show that bonded joints remain a critical failure point for e-textile interconnections. The reported results show that rigid epoxied joints produce the best results but ultimately result in the cracking of the conductive patterns. Flexible connectors with better adhesion than currently available pressure-bonded conductive films are still required to minimise the stiffness mismatch created by epoxy joints. It is recommended that the durability of the printed e-textiles could be improved by minimizing the number of bonded joints where possible during e-textile design and manufacture; this can be achieved by integrating more of the electronics into or onto the textile, thus reducing the need for interconnections.

Author Contributions: These authors equally contributed to this work. All authors have read and agreed to the published version of the manuscript.

Funding: This work was funded by the WEARPLEX project with the grant agreement ID 825339 under the EU Horizon 2020 funding—ICT-02-2018: www.wearplex.soton.ac.uk (accessed on 22 April 2022).

Data Availability Statement: The data for this paper can be found at: doi.org/10.5258/SOTON/D2192.

Conflicts of Interest: The authors declare no conflict of interest.

References

1. Paul, G.; Torah, R.; Beeby, S.; Tudor, J. A printed, dry electrode Frank configuration vest for ambulatory vectorcardiographic monitoring. *Smart Mater. Struct.* **2017**, *26*, 025029. [CrossRef]
2. Hardy, D.A.; Moneta, A.; Sakalyte, V.; Connolly, L.; Shahidi, A.; Hughes-Riley, T. Engineering a costume for performance using illuminated LED-yarns. *Fibers* **2018**, *6*, 35. [CrossRef]
3. Almusallam, A.; Torah, R.N.; Zhu, D.; Tudor, M.J.; Beeby, S.P. Screen-printed piezoelectric shoe-insoleenergy harvester using an improved flexible PZT-polymer composites. *J. Phys. Conf. Ser.* **2013**, *476*, 012108. [CrossRef]
4. Komolafe, A.; Zaghari, B.; Torah, R.; Weddell, A.; Khanbareh, H.; Tsikriteas, Z.M.; Beeby, S. E-textile Technology Review—From Materials to Application. *IEEE Access* **2021**, *9*, 97152–97179. [CrossRef]
5. Koshi, T.; Nomura, K.I.; Yoshida, M. Measurement and analysis on failure lifetime of serpentine interconnects for e-textiles under cyclic large deformation. *Flex. Print. Electron.* **2021**, *6*, 025003. [CrossRef]
6. Komolafe, A. Reliability and Interconnections for Printed Circuits on Fabrics. Ph.D. Thesis, University of Southampton, Southampton, UK, 2016.
7. Merilampi, S.; Laine-Ma, T.; Ruuskanen, P. The characterization of electrically conductive silver ink patterns on flexible substrates. *Microelectron. Reliab.* **2009**, *49*, 782–790. [CrossRef]
8. Jin, H.; Matsuhisa, N.; Lee, S.; Abbas, M.; Yokota, T.; Someya, T. Enhancing the performance of stretchable conductors for e-textiles by controlled ink permeation. *Adv. Mater.* **2017**, *29*, 1605848. [CrossRef] [PubMed]
9. Szalapak, J.; Scenev, V.; Janczak, D.; Werft, L.; Rotzler, S.; Jakubowska, M.; Schneider-Ramelow, M. Washable, Low-Temperature Cured Joints for Textile-Based Electronics. *Electronics* **2021**, *10*, 2749. [CrossRef]
10. Komolafe, A.; Torah, R. Effect of textile primer layer on screen printed conductors for e-textiles. In Proceedings of the 2021 IEEE International Conference on Flexible and Printable Sensors and Systems (FLEPS), Manchester, UK, 20–23 June 2021; pp. 1–4.
11. Komolafe, A.O.; Nunes-Matos, H.; Glanc-Gostkiewicz, M.; Torah, R.N. Evaluating the effect of textile material and structure for printable and wearable e-textiles. *IEEE Sens. J.* **2021**, *21*, 18263–18270. [CrossRef]
12. Smart Fabric Inks Ltd. Smart Fabric Technology. Available online: http://www.fabinks.com/ (accessed on 3 December 2021).
13. Komolafe, A.; Torah, R.; Wei, Y.; Nunes-Matos, H.; Li, M.; Hardy, D.; Beeby, S. Integrating flexible filament circuits for e-textile applications. *Adv. Mater. Technol.* **2019**, *4*, 1900176. [CrossRef]

Proceeding Paper

Optimization of Knitted Structures for E-Textiles Applications †

Sultan Ullah, Khubab Shaker and **Syed Talha Ali Hamdani** *

School of Engineering and Technology, National Textile University, Faisalabad 37610, Pakistan; sultankhattak@ntu.edu.pk (S.U.); khubab@ntu.edu.pk (K.S.)
* Correspondence: hamdani.talha@ntu.edu.pk; Tel.: +92-41-9230081-82 (ext. 295)
† Presented at the 3rd International Conference on the Challenges, Opportunities, Innovations and Applications in Electronic Textiles (E-Textiles 2021), Manchester, UK, 3–4 November 2021.

Abstract: The findings of this research attempt to evaluate the electrical and compression features of electrically conductive yarns (ECY) as well as the structure of sensor systems, such as single jersey and double jersey knit designs, for healthcare applications and wearing technologies. The tensile properties and electrical properties of conductive yarns were optimized basis of the findings. Owing to the knit-tuck stitches arrangement, which gives density to the fabric, the double lacoste, popcorn, and milano ribs were proven to have adequate compressive resilience. The developed knitted structures kinds of sensors were noticed and may easily be applied to global smart socks manufacture as well as other technologies.

Keywords: silver-coated yarns; pressure sensor; electrical properties; compression properties; e-textiles

Citation: Ullah, S.; Shaker, K.; Hamdani, S.T.A. Optimization of Knitted Structures for E-Textiles Applications. *Eng. Proc.* **2022**, *15*, 18. https://doi.org/10.3390/engproc2022015018

Academic Editors: Steve Beeby, Kai Yang and Russel Torah

Published: 20 May 2022

Publisher's Note: MDPI stays neutral with regard to jurisdictional claims in published maps and institutional affiliations.

1. Introduction

By recognizing and responding to a sensory input, smart textiles response to the environment with various components of electronics in the form of yarns or textiles [1] integration of various patterns and composition of woven, non-woven, and knitted structures [2].

2. Results and Discussion

The conductive yarns 280D-FDY and SPFX25070-FX were selected based on the results shown in Figure 1a, and the electrical characteristics of the conductive yarns were seen in Figure 1b. The Kawabata evaluation system (KES FB-03) had been used to test pressure sensors such as double lacoste, popcorn, and milano rib compression properties, with the findings reported in Table 1. Figure 2 and Table 2 show the pressure electrical resistance curve for an optimized knitted structure. The decrease in electrical resistance when subjected to varying loads indicates well for the insertion of such designs into socks for counting calories and other health-related applications in the field of e-textiles. Furthermore, the sensors utilized in health monitoring systems are extremely adaptable, allowing for a natural interface with the human body [3].

Table 1. Compression results for optimized knitted pressure sensors.

Compression Characteristics	Double Lacoste	Popcorn	Milano Rib
Linearity of compression (LC)	0.560 ± 0.036	0.570 ± 0.026	0.550 ± 0.050
Work of compression (WC) gf. cm/cm^2	1.467 ± 0.108	1.850 ± 0.050	2.157 ± 0.137
The resilience of compression (RC) %	44.000 ± 1.375	45.557 ± 0.407	49.55 ± 0.918
Thickness at the max load (To) mm	3.550 ± 0.050	3.490 ± 0.066	4.666 ± 0.015
Thickness at the pressure 0.5 g. f/cm^2 (Tm) mm	2.447 ± 0.186	1.550 ± 0.050	1.550 ± 0.05

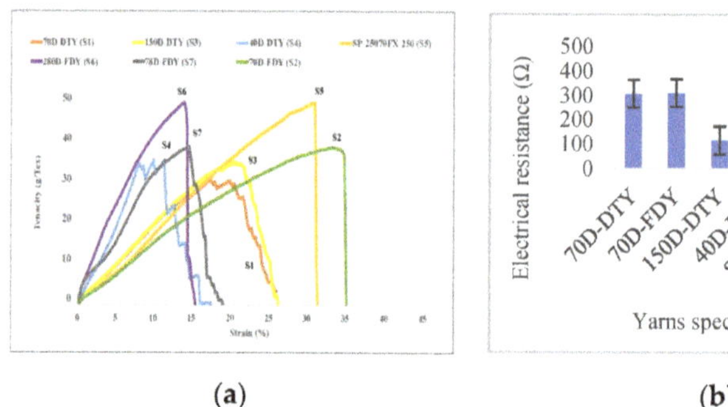

(a)   **(b)**

Figure 1. (**a**) Tenacity-strain curve for conductive yarns; (**b**) electrical resistance (Ω) of conductive yarns.

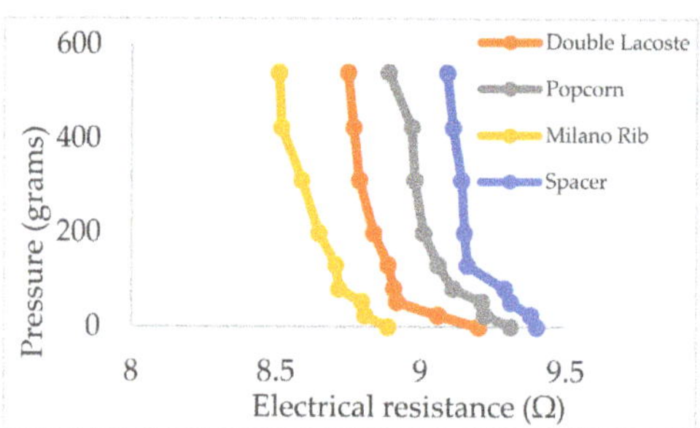

Figure 2. Pressure-electrical resistance curve for an optimized knitted pressure sensor.

Table 2. Electrical resistance measurement for the selected knitted structures to identify the pressure sensing properties.

Sr. #.	Pressure (Grams)	Double Lacoste Resistance (Ω)	Popcorn Resistance (Ω)	Milano Rib Resistance (Ω)	Spacer Resistance (Ω)
1	0	9.20 ± 0.21	9.31 ± 0.32	8.89 ± 0.39	9.40 ± 0.41
2	25	9.06 ± 0.23	9.22 ± 0.29	8.81 ± 0.32	9.38 ± 0.41
3	53	8.92 ± 0.23	9.21 ± 0.22	8.80 ± 0.29	9.31 ± 0.29
4	81	8.91 ± 0.25	9.11 ± 0.22	8.72 ± 0.31	9.29 ± 0.28
5	131	8.89 ± 0.28	9.06 ± 0.19	8.71 ± 0.29	9.16 ± 0.31
6	199	8.84 ± 0.22	9.01 ± 0.23	8.65 ± 0.33	9.15 ± 0.33
7	311	8.79 ± 0.23	8.98 ± 0.28	8.59 ± 0.35	9.14 ± 0.41
8	424	8.77 ± 0.25	8.97 ± 0.26	8.52 ± 0.41	9.11 ± 0.39
9	540	8.75 ± 0.22	8.89 ± 0.33	8.51 ± 0.29	9.09 ± 0.36

3. Conclusions

Based on their compressional properties, this research work sought to select/optimize the best practicable knitted structures from both single and double jersey knitted structures. Double lacoste and popcorn were reported to have better compressional behavior in case of single jersey knitted structures. The only structure in a double jersey is the milano rib, which offers a higher compressive value due to the structure design and the best pressure sensing properties. Smart textiles are seen as the industry's future, with numerous new items being developed in various stages of life in response to demand [4]. Miniaturization of health monitoring systems is progressing to manage complicated computing and efficient information sensing [5]. Several studies [6,7] have been done to improve the sensing characteristics of textile constructions made from electrically conductive yarns.

Author Contributions: S.U. contributed to original draft preparation, methodology and investigation. K.S. helped in supervision and formal analysis. S.T.A.H. contributed to conceptualization supervision, project administration and funding acquisition. All authors have read and agreed to the published version of the manuscript.

Funding: This research was funded by Higher Education Commission of Pakistan, grant number TDF-03/056.

Institutional Review Board Statement: Not applicable.

Informed Consent Statement: This study was conducted to measure the calories burnt during human activities like running, jumping, walking etc., and to get the actual quantities consumed. The purpose of this study was to manufacture a textile-based structure for E-Textiles applications along with good features of comfortability and breathability.

Data Availability Statement: Not applicable.

Conflicts of Interest: The authors declare no conflict of interest.

References

1. Zhao, J.; Fu, Y.; Xiao, Y.; Dong, Y.; Wang, X.; Lin, L. A Naturally Integrated Smart Textile for Wearable Electronics Applications. *Adv. Mater. Technol.* **2020**, *5*, 1900781. [CrossRef]
2. Stoppa, M.; Chiolerio, A. Wearable electronics and smart textiles: A critical review. *Sensors* **2014**, *14*, 11957–11992. [CrossRef] [PubMed]
3. Gao, W.; Ota, H.; Kiriya, D.; Takei, K.; Javey, A. Flexible Electronics Toward Wearable Sensing. *Acc. Chem. Res.* **2019**, *52*, 523–533. [CrossRef] [PubMed]
4. Syduzzaman, M.D.; Patwary, S.U.; Farhana, K.; Ahmed, S. Smart Textiles and Nano-Technology: A General Overview. *J. Text. Sci. Eng.* **2015**, *5*, 1000181.
5. Mahmud, M.S.; Wang, H.; Esfar-E-Alam, A.M.; Fang, H. A Wireless Health Monitoring System Using Mobile Phone Accessories. *IEEE Internet Things J.* **2017**, *4*, 2009–2018. [CrossRef]
6. Kang, B.G.; Hannawald, J.; Brameshuber, W. Electrical resistance measurement for damage analysis of carbon yarns. *Mater. Struct.* **2011**, *44*, 1113–1122. [CrossRef]
7. Goldfeld, Y.; Perry, G. Electrical characterization of smart sensory system using carbon based textile reinforced concrete for leakage detection. *Mater. Struct.* **2018**, *51*, 170. [CrossRef]

engineering proceedings

Proceeding Paper

Textile Manufacturing Compatible Triboelectric Nanogenerator with Alternating Positive and Negative Woven Structure †

Watcharapong Paosangthong *, Mahmoud Wagih , Russel Torah and Steve Beeby

Smart Electronic Materials and Systems Research Group, Electronics and Computer Science, University of Southampton, Southampton SO17 1BJ, UK; mahm1g15@soton.ac.uk (M.W.); rnt@ecs.soton.ac.uk (R.T.); spb@ecs.soton.ac.uk (S.B.)
* Correspondence: wp1y15@soton.ac.uk
† Presented at the 3rd International Conference on the Challenges, Opportunities, Innovations and Applications in Electronic Textiles (E-Textiles 2021), Manchester, UK, 3–4 November 2021.

Abstract: This paper reports the novel design of a textile-based triboelectric nanogenerator (TENG) with alternate woven strips of positive and negative triboelectric material operating in freestanding triboelectric-layer mode (woven TENG). It was fabricated using processes that are compatible with standard textile manufacturing, including plain weaving and doctor blading. In contrast to the conventional grating structure TENGs, which can operate only in one moving direction, this new design allows the woven TENG to operate in all planar directions. The implementation of the positive and negative triboelectric material also significantly improves the performance of the woven TENG compared to the conventional all-direction TENGs.

Keywords: e-textile; triboelectric nanogenerator; woven structure

Citation: Paosangthong, W.; Wagih, M.; Torah, R.; Beeby, S. Textile Manufacturing Compatible Triboelectric Nanogenerator with Alternating Positive and Negative Woven Structure. *Eng. Proc.* **2022**, *15*, 19. https://doi.org/10.3390/engproc2022015019

Academic Editor: Kai Yang

Published: 2 June 2022

Publisher's Note: MDPI stays neutral with regard to jurisdictional claims in published maps and institutional affiliations.

1. Introduction

Despite intensive research and development in wearable and portable electronics, most of these types of devices are still powered by batteries, which require persistent recharging and replacement. Introducing a wearable self-charging power system using an energy harvester to scavenge energy from human body movements is one of the most effective ways to solve this problem. The triboelectric nanogenerator (TENG) is one of the most promising candidates for powering these devices. They can efficiently convert kinetic energy occurring after or during frictional contact between two dissimilar materials into electricity based on contact electrification and electrostatic induction effects [1]. They are determined to be highly suitable for powering wearable devices and electronic textiles (e-textiles) due to their properties, including low weight, flexibility, washability, high efficiency, and biocompatibility [2–4].

This paper proposes a novel structure of textile-based TENG with alternate woven strips of positive and negative triboelectric material operating in sliding freestanding triboelectric-layer mode, defined as a woven TENG. They can harvest energy from all arbitrary sliding directions. Whereas most of the TENGs with this ability are made of rigid materials and thus unsuitable for e-textiles [5–7], this woven TENG is fabricated using processes which are compatible with standard textile manufacturing, including plain weaving and doctor blading. Moreover, the introduction of the positive and negative triboelectric material considerably enhances the power output of the device.

2. Device and Working Principle

The device is composed of a woven upper substrate and woven electrodes, as shown in Figure 1a. The photograph of the upper substrate is illustrated in Figure 1b. It is fabricated by plain-weaving strips of adhesive nylon fabric (Hemline) and PTFE-coated fibreglass fabric (Xinghoo) forming a checker-like structure. The adhesive layer on the back side of the nylon fabric and the PTFE coating layer bond their fibers together and prevent the fabrics from fraying when they are cut into strips. The nylon fabric and the PTFE fabric were selected as the positive triboelectric material and the negative triboelectric material since they are some of the most positive and negative materials in the triboelectric series, respectively. Additionally, they are available in a commercial fabric form and have good properties for implementation in wearable devices, such as flexibility, lightweight, biocompatibility, and washability.

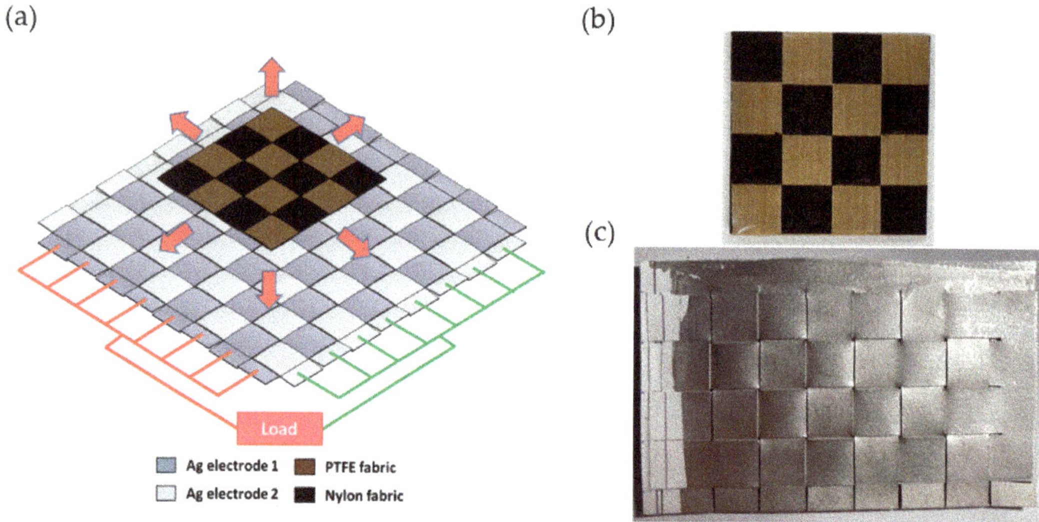

Figure 1. (**a**) Schematic illustration and photographs of woven TENG for $N = 4$ comprising (**b**) an upper substrate with woven strips of nylon fabric and PTFE fabric and (**c**) lower woven Ag electrodes.

The photograph of the lower electrodes is illustrated in Figure 1c. It was fabricated by doctor blade coating Ag ink (Fabinks TC-C4001) on a PVC-coated polyester fabric, defined as PVC fabric (VALMEX FR 7546). Two pieces of PVC fabric were firstly pre-baked at 120 °C for 15 min in a Carbolite box oven to eliminate outgassing from the PVC layer. After that, a doctor blade was used to coat the PVC fabric with a 100 μm-thick Ag ink, and was cured at a temperature of 120 °C for 15 min. The Ag-coated PVC fabrics were then cut into strips of the same size with a 1 mm gap between each strip, leaving one end uncut. Finally, the two Ag-coated PVC fabrics were woven together, forming the lower woven electrodes. The uncut ends enable the warp strips and weft strips to be electrically connected forming separate warp and weft electrodes. The 1 mm gap ensures that the warp and weft electrode are electrically isolated after weaving. The upper substrate and the electrodes were produced with various numbers of elements (N) on each side with N equalling 2, 4, 6, and 8 with strip widths of 4 cm, 2 cm, 1.33 cm, and 1 cm, respectively. The total size of the upper substrate was fixed at 8 cm × 8 cm. Alternatively, the electrodes can be fabricated by attaching pre-cut adhesive vinyl stencil strips with a width of 1 mm to the PVC fabric before doctor-blading it with Ag ink. After removing the vinyl stencil strips, the 1 mm uncoated gaps were formed between the electrode strips. The advantage of this method is that it can prevent the electrodes from electrical short circuits.

With regards to the Ag electrodes, when the positive (nylon fabric) and negative triboelectric materials (PTFE fabric) are brought into contact with the Ag electrodes, positive and negative charges build up at their surface, respectively, and at the same time, the same amount of charge with the opposite polarity can be transferred to the electrodes due to the triboelectric effect. Through a sliding movement of the woven triboelectric materials across the woven electrodes, the electrons were induced and transferred between the weft and the warp electrode due to the potential difference at the electrodes caused by the shielding effect and the difference in distance between the triboelectric materials and the top and bottom electrode. Through this periodic movement, multiple alternating currents were generated.

3. Experimental Results

The experiments were performed using a belt-driven linear actuator at a mechanical oscillation of 2 Hz, a travel distance of 40 mm, a contact force of 5 N, a humidity of 25% RH, and a temperature of 25 °C. Figure 2a shows the root mean square (RMS) values of the open-circuit voltage (V_{OC}) and the short-circuit current (I_{SC}) of the woven TENG for the different element numbers. The V_{OC} slightly decreases as the element number increases due to a growth in the capacitance of the electrodes with smaller strips, whereas the I_{SC} shows a strong increasing trend due to a reduction in the travel time of the charge transfer. The dependence of the RMS voltage, RMS current, and RMS power on the external load resistance of the woven TENG with $N = 8$ is revealed in Figure 2b. The RMS voltage and RMS current peak at 62.9 V and 1.77 µA, respectively. The RMS power reaches a maximum of 34.77 µW at a load resistance of 50 MΩ, corresponding to a maximum power density of 5.43 mW/m^2.

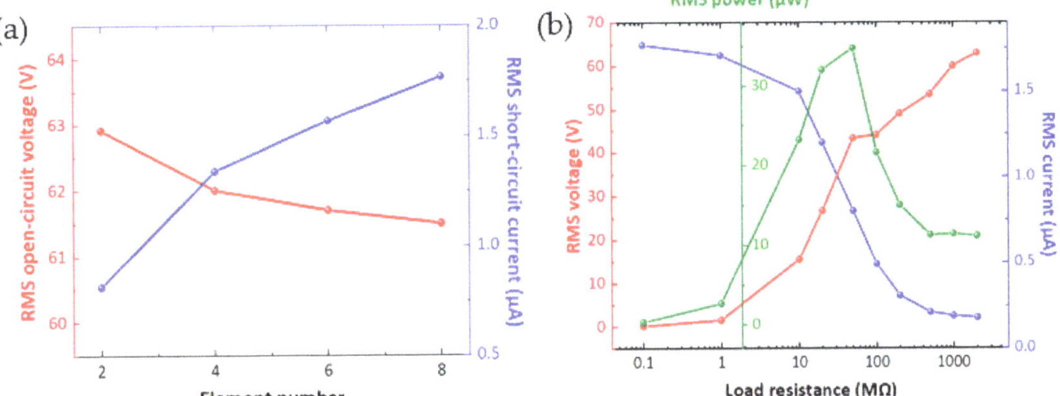

Figure 2. (**a**) RMS values of V_{OC} and I_{SC} for different *N*. (**b**) Dependence outputs of the woven TENG with $N = 8$ on the external load resistance.

The experimental RMS V_{OC} and RMS I_{SC}, as functions of the sliding direction with respect to the weft electrode alignment, are plotted in Figure 3a. They are compared with the percentage of the maximum output calculated from the ratio of the overlapping areas between the upper substrate and the electrodes at each sliding direction. The maximum outputs are obtained when the upper substrate slides are parallel or perpendicular to the weft electrode, and the minimum outputs of 50% of the maximum outputs are produced when the upper substrate moves diagonally with respect to the weft electrode (45 degrees). Figure 3b shows the transient V_{OC} for the woven TENG with PTFE and nylon fabric ($N = 4$), a checker-like structured TENG with PTFE fabric and a checker-like structured TENG with nylon fabric. The V_{OC} of the woven TENG rises by 1.3 times and 2.6 times compared to the TENG with only PTFE fabric and nylon fabric, respectively.

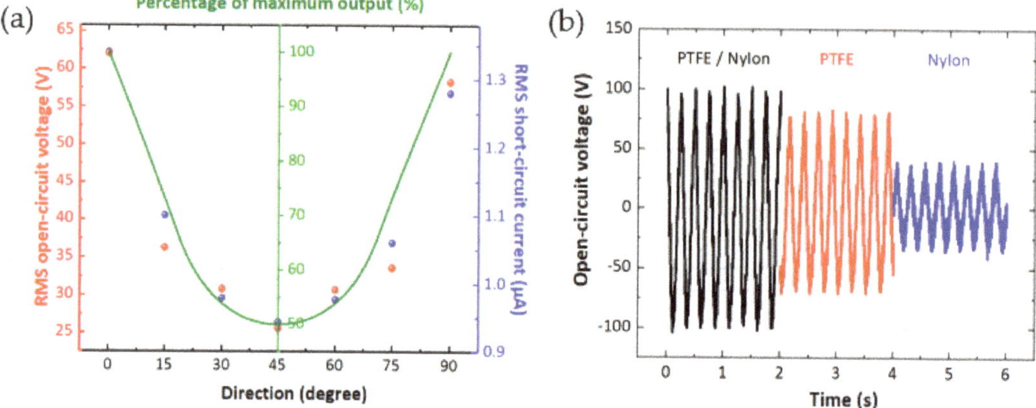

Figure 3. (**a**) Calculated percentage of maximum output compared with experimental RMS V_{OC} and RMS I_{SC} as a function of sliding direction with respect to the weft electrode alignment. (**b**) Transient V_{OC} of different types of TENGs.

To demonstrate a possible use of the woven TENG in wearable electronics, the woven TENG ($N = 8$) was embedded into a lab coat. The energy is generated from the relative movement between the arm and the torso during walking and running. The output of the woven TENG was used to charge capacitors with different capacitances. The capacitors can be charged to a useful voltage for wearable electronics in a short period. For example, it takes 4 s and 15 s to charge a 1 µF capacitor to 3 V for running and walking, respectively. The output was also used to drive a wearable night-time warning indicator for pedestrians (Figure 4a), a digital watch (Figure 4b), and a Bluetooth transceiver. As a demonstration of a sensing device (Figure 4c), the voltage peaks of the woven TENG was detected and applied for step counting through arm motion (pedometer). Additionally, the device also shows good washability and durability. It can withstand 40,000 operating cycles and 5 washes without significant change in the output.

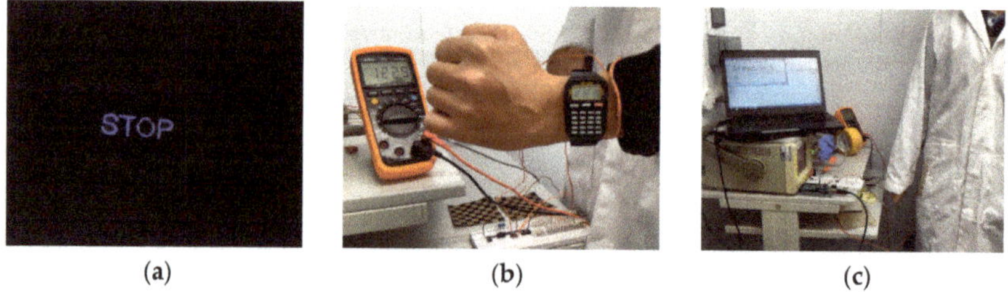

(**a**) (**b**) (**c**)

Figure 4. Photograph of (**a**) night-time warning indicator, (**b**) digital watch, and (**c**) pedometer.

4. Conclusions

A novel textile-based triboelectric generator was successfully fabricated using processes that are compatible with standard textile manufacturing, including plain weaving and doctor blade coating, defined as woven TENG. The woven TENG comprises a woven upper substrate with alternating positive (nylon fabric) and negative triboelectric material (PTFE fabric) and woven Ag electrodes with matching structure. The woven TENG can generate energy in all sliding directions, demonstrating the significant performance improvement compared to the TENGs with single triboelectric materials. This paper has demonstrated the applications for the woven TENG, including a wearable night-time warning indicator, a digital watch, a Bluetooth transceiver, and a pedometer.

Author Contributions: Conceptualization, methodology, writing, editing W.P.; applications, W.P. and M.W.; supervision, R.T. and S.B. All authors have read and agreed to the published version of the manuscript.

Funding: This research was funded by the Engineering and Physical Sciences Research Council (EPSRC) with grant reference EP/P010164/1.

Institutional Review Board Statement: Not applicable.

Informed Consent Statement: Not applicable.

Data Availability Statement: Not applicable.

Conflicts of Interest: The authors declare no conflict of interest. The funders had no role in the design of the study; in the collection, analyses, or interpretation of data; in the writing of the manuscript, or in the decision to publish the results.

References

1. Fan, F.-R.; Tian, Z.-Q.; Lin Wang, Z. Flexible triboelectric generator. *Nano Energy* **2012**, *1*, 328–334. [CrossRef]
2. Paosangthong, W.; Torah, R.; Beeby, S. Recent progress on textile-based triboelectric nanogenerators. *Nano Energy* **2019**, *55*, 401–423. [CrossRef]
3. Paosangthong, W.; Wagih, M.; Torah, R.; Beeby, S. Textile Manufacturing Compatible Triboelectric Nanogenerator with Alternating Positive and Negative Freestanding Grating Structure. *Proceedings* **2020**, *32*, 23.
4. Paosangthong, W.; Wagih, M.; Torah, R.; Beeby, S. Textile-based triboelectric nanogenerator with alternating positive and negative freestanding grating structure. *Nano Energy* **2019**, *66*, 104148. [CrossRef]
5. Guo, H.; Leng, Q.; He, X.; Wang, M.; Chen, J.; Hu, C.; Xi, Y. A triboelectric generator based on checker-like interdigital electrodes with a sandwiched PET thin film for harvesting sliding energy in all directions. *Adv. Energy Mater.* **2015**, *5*, 1400790. [CrossRef]
6. Li, X.; Xu, C.; Wang, C.; Shao, J.; Chen, X.; Wang, C.; Tian, H.; Wang, Y.; Yang, Q.; Wang, L.; et al. Improved triboelectrification effect by bendable and slidable fish-scale-like microstructures. *Nano Energy* **2017**, *40*, 646–654. [CrossRef]
7. Xia, X.; Liu, G.; Guo, H.; Leng, Q.; Hu, C.; Xi, Y. Honeycomb-like three electrodes based triboelectric generator for harvesting energy in full space and as a self-powered vibration alertor. *Nano Energy* **2015**, *15*, 766–775. [CrossRef]

Proceeding Paper

Flexible Water-Activated Battery on a Polyester–Cotton Textile [†]

Sheng Yong [1,*], **Nick Hillier** [1,2] **and Stephen Beeby** [1]

1 Smart Electronic Materials & System Research Group, Electronics and Computer Science, University of Southampton, Southampton SO17 1BJ, UK; nh3g09@soton.ac.uk (N.H.); spb@ecs.soton.ac.uk (S.B.)
2 Energy Storage and Its Applications Centre of Doctoral Training, University of Southampton, Southampton SO17 1BJ, UK
* Correspondence: sy1v16@soton.ac.uk; Tel.: +44-2380523119
† Presented at the 3rd International Conference on the Challenges, Opportunities, Innovations and Applications in Electronic Textiles (E-Textiles 2021), Manchester, UK, 3–4 November 2021.

Abstract: This work presents a simple, scalable and flexible water activated primary battery, fabricated on top of a textile substrate. The textile-based battery was fabricated with inexpensive screen-printed materials to form the functional battery electrodes, novel polymer separator and buffer layers. Upon activating the device with water, the battery demonstrated an areal capacity of 88 μAh·cm^{-2} between 1 and 0.6 V.

Keywords: e-textile; flexible battery; water-activated battery

Citation: Yong, S.; Hillier, N.; Beeby, S. Flexible Water-Activated Battery on a Polyester–Cotton Textile. *Eng. Proc.* **2022**, *15*, 20. https://doi.org/10.3390/engproc2022015020

Academic Editors: Steve Beeby, Kai Yang and Russel Torah

Published: 10 June 2022

Publisher's Note: MDPI stays neutral with regard to jurisdictional claims in published maps and institutional affiliations.

1. Introduction

Electronic textiles (e-textiles) are the combination of electrical devices with flexible substrate such as woven, knitted or non-woven textiles. Alongside the electrical functionality of e-textile systems, the power supply of these electronics is the key challenge for their application in real world scenarios [1]. Integrating flexible energy storage/supply elements (such as batteries) within textiles is seen as a key technology to overcome this challenge, and has rightfully seen a significant increase in research interest [2].

Water-activated batteries are a single use electrical energy storage device. In comparison to ordinary primary batteries, they do not contain the fluidic electrolyte and need to be activated with water to power the connected electrical system. This type of battery demonstrates distinct advantages, such as the following: light weight, reliable, flexible, long shelf life prior to activation and high power and energy density [3]. It has been used in textile devices such as life jackets, medical sensing diapers and military garments [4]. Historically, textile water-activated batteries have been fabricated with multiple textile layers that are coated with different functional materials. In 2015, Liu, et al. [5] reported the first water-activated battery with three textile layers with different functional materials and two metal layers for the current collectors and anode/cathode material. This device achieved a voltage of 1.3 V and could power an LED for 30 min when two were connected in series. The battery cell was sealed with double sided tape to prevent the battery from short circuiting and to ensure all material layers were in contact with each other. In 2018, Vilkhuet et al. [6] implemented a full water-activated battery with battery electrodes printed on the same side of the textile. After wetting the battery with a highly alkaline solution with sodium hydroxide (6 M), the unpackaged battery cell achieved an initial voltage of 1.46 V, with an approximate areal capacity of 400 μAh·cm^{-2} above 0.2 V. In 2020, Yi et al. [7] presented a flexible, encapsulated and printable water activated primary battery. The battery/galvanic cell was fabricated from two pieces of modified cotton textile to form the anode and porous separator, with an aluminium metal foil as the cathode. Upon activation, the battery achieved an open circuit voltage of 1.32 V and an areal capacity of 166.8 μAh·cm^{-2} above 0.8 V. These examples demonstrate the capability of fabricated water-activated batteries within textile material. However, the use of multiple functional textile or metal layers increased the device thickness and mechanical inflexibility, introducing extra encapsulation challenges for real-world power source units in e-textile systems. In addition, the requirement of highly concentrated alkaline solution as the battery activation agent (as in the work of Vilkhuet et al.) leads to extra encapsulation difficulties to stop solution leakage, accompanied by environmental hazards. Therefore, these designs were not suitable for real-word e-textile applications.

This paper reports an approach for fabricating a water-activated battery with one textile and one metal/polymer film. The battery's anode was a flexible zinc polymer film prepared via screen printing, whilst the battery's cathode was a flexible and screen-printed silver polymer layer on a polyester cotton textile. The separator and buffer layers were implemented on top of the battery's anode and cathode with doctor blading or screen printing followed by a phase inversion process. The battery was tested after it had been activated with water to study its discharge performance.

2. Material and Methods

The proposed water activated primary battery was fabricated on top of a polyester-cotton textile of thickness of ~200 µm using solution-based processes and inexpensive materials. Figure 1 shows the proposed structure of the primary battery. Firstly, silver paste (Fabinks TC-C4001) was screen printed onto the polyester–cotton textile, where the silver paste soaked through the textile, forming the current collector and cathode of the device (Figure 2a). Then a gel paste (containing 2 M silver nitrate ($AgNO_3$) and polyvinyl alcohol (PVA) (1:10 by wt.%)) was screen printed on top of the silver-coated textile. The textile was then sonicated for 10 min in isopropyl alcohol until a white porous film formed on top of the silver-coated textile. The coated textile was then dried under a 25 mbar vacuum at room temperature (23 °C) for 30 min. The separator layer was fabricated via doctor blade coating with a gel solution containing 6 M sodium nitrate ($NaNO_3$) and polyacrylamide (PAM) (1:20 by wt%) onto the textile cathode. It was sonicated for 30 min with acetone and completely dried under vacuum at room temperature (23 °C) for 2 h (Figure 2b). The anode of the water-activated primary battery was fabricated with the same processes as the cathode textile layer: A zinc paste containing zinc metal powder and polymer binder polyethylene-vinyl acetate (5:95 by wt%) was screen printed onto a teflon sheet and peeled off after complete evaporation of the solvent under vacuum at room temperature (23 °C), which formed the current collector and anode of the device (Figure 2c). Then, a gel paste containing 3 M zinc chloride ($ZnCl_2$) and PVA (1:10 by wt%) was screen printed on top of a flexible zinc anode film, before undergoing the same sonication and drying process (Figure 2d).

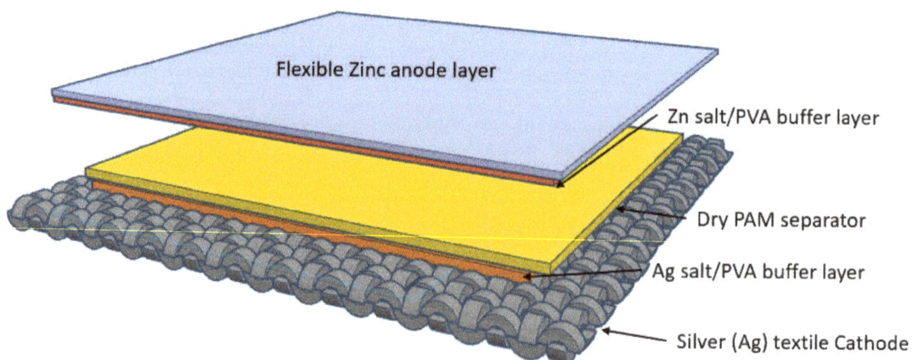

Figure 1. Water-activated battery schematic.

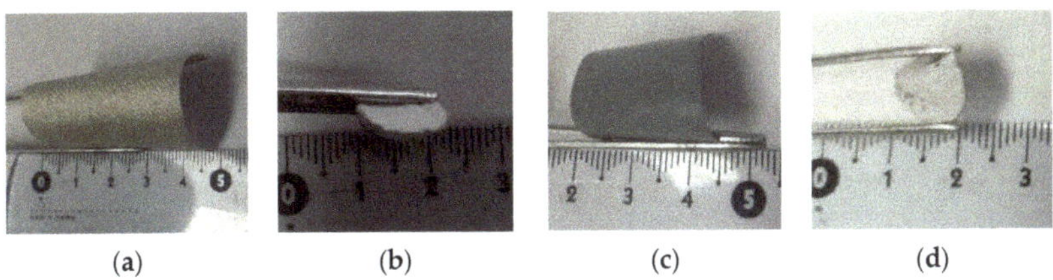

(a)   (b)   (c)   (d)

Figure 2. (**a**) Flexible silver textile. (**b**) Silver textile with separator. (**c**) Flexible zinc film. (**d**) Zinc film with buffer film.

Figure 3a shows the test setup for the battery performance testing. The proposed water-activated primary battery was made with one piece of zinc anode film and silver textile cathode/separator each; these layers were punched into circular shapes with a diameter of 1 cm and encapsulated within a Swagelok tube fitting for the discharge performance test. The total thickness of the battery, including the zinc anode, textile cathode, two salt buffer layers and PAM separator layer, was ~2 mm. Figure 3b shows the test setup for the encapsulated battery. Two pieces of hotmelt polymer film were used to encapsulate the water-activated primary battery. An air hole with a diameter of 0.3 cm was punched in the centre of cathode textile, PAM separator and one of the encapsulation films to allow the device to be activated by a deionised (DI) water droplet before testing.

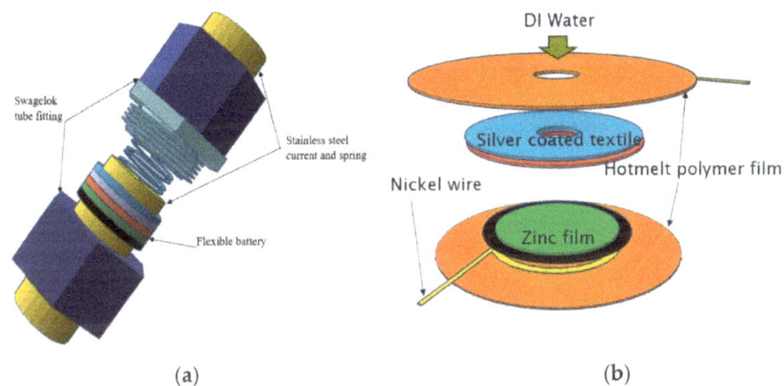

(a)   (b)

Figure 3. (**a**) Battery test in Swagelok tube fitting. (**b**) Encapsulated battery test configuration.

3. Results

The discharge performance of the textile water-activated batteries were characterised with an Autolab Pgsatat101 (Metrohm Autolab, Utrecht, The Netherlands). DI water (0.1 mL) was used to activate the battery before being put into the test tube fitting or through the air hole for the encapsulated device. The discharge current was set to be $0.5\ \text{mA}\cdot\text{cm}^{-2}$ or 0.395 mA.

The water-activated battery in the Swagelok test configuration (Figure 4a) achieved an initial voltage above 1 V, with an areal capacity of $88\ \mu\text{Ah}\cdot\text{cm}^{-2}$ or energy density of $144\ \mu\text{Wh}\cdot\text{cm}^{-2}$ above 0.6 V. The results of the water-activated battery on textile encapsulated with polymer films (Figure 4b) demonstrates that the device voltage rose from 0.75 to 0.9 V for 100 s after water activation, with the total capacity of this type of device reaching $55\ \mu\text{Ah}\cdot\text{cm}^{-2}$ ($97.2\ \mu\text{Wh}\cdot\text{cm}^{-2}$) above 0.6 V. In both types of device, the polymer separator layer was wetted by the water, which turned the separator into a porous gel membrane that contained dissolved $NaNO_3$. This membrane acted as the salt bridge, allowing ion transfer between the cathode and anode whilst preventing any electrical short circuits. With the polymer film encapsulated device, the wetting process took longer than that of the battery in the Swagelok, which explains the voltage increase before discharge.

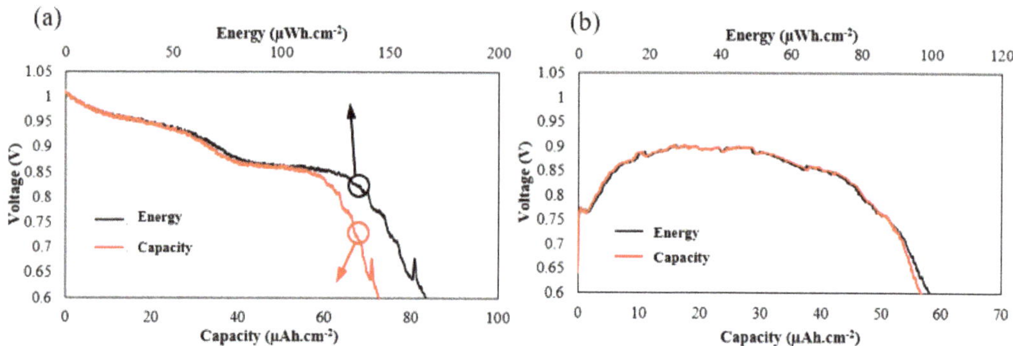

Figure 4. Discharge test (discharge current = 0.5 mA·cm^{-2}) of the water-activated battery on textile. (**a**) In tube fitting. (**b**) Encapsulated with polymer films.

4. Discussion and Concussions

This paper demonstrates a simple and straightforward fabrication method of a textile water-activated battery with a single piece of textile and metal/polymer layer, with both layers being flexible and scalable. The printed and phased inverted separator layer on top of the textile cathode shows a promising way of fabricating a porous gel membrane that can act as salt bridge in future batteries. The battery tested in a Swagelok achieved an areal capacity of 88 μAh·cm^{-2} or energy density of 144 μWh·cm^{-2}, whilst the polymer film encapsulated battery demonstrated an areal capacity of 55 μAh·cm^{-2} or energy density of 97.2 μWh·cm^{-2}. The proposed battery is flexible and scalable, and can be easily integrated into e-textile systems as an emergency electrical power supply. Future works will include optimising the device structure and material formulation for better energy storage performance.

Author Contributions: All authors contributed to the research equally. All authors have read and agreed to the published version of the manuscript.

Funding: This research was funded and supported by the United States Army, with Contract No. W911NF2010324 and EPSRC (grants EP/P010164/1 and EP/L016818/1).

Institutional Review Board Statement: Not applicable.

Informed Consent Statement: Not applicable.

Data Availability Statement: The data presented in this study are available on request from the corresponding author.

Acknowledgments: This work was also supported by the Royal Academy of Engineering under the Chairs in Emerging Technologies scheme.

Conflicts of Interest: Beeby is the director of Smart Fabric Inks.

References

1. Vališevskis, A.; Briedis, U.; Carvalho, M. Development of flexible textile aluminium-air battery prototype. *Mater. Renew. Sustain. Energy* **2021**, *10*, 1–6. [CrossRef]
2. Mauriello, M.; Gubbels, M.; Froehlich, J.E. Social fabric fitness: The design and evaluation of wearable E-textile displays to support grouprunning. In Proceedings of the SIGCHI Conferences on Human Factors in Computing Systems, Toronto, ON, Canada, 26 April–1 May 2014; pp. 2833–2842. [CrossRef]
3. Nguyen, D.; Ito, Y.; Taguchi, K. A water-activated paper-based battery based on activated carbon powder anode and CuCl$_2$/CNT cathode. *Energy Rep.* **2020**, *6*, 215–219. [CrossRef]
4. Reddy, T.B. *Handbook of Batteries*, 3rd ed.; McGraw-Hill: New York, NY, USA, 2001; Chapter 16.
5. Liu, X.; Lillehoj, P.B. A liquid-activated textile battery for wearable biosensors. *J. Physics Conf. Ser.* **2015**, *660*, 012063. [CrossRef]

6. Vilkhu, R.; Thio, W.; Das Ghatak, P.; Sen, C.K.; Co, A.; Kiourti, C. Power generation for wearable electronics: Designing electrochemical storage on fabrics. *IEEE Access* **2018**, *6*, 28945–28950. [CrossRef] [PubMed]
7. Li, Y.; Yong, S.; Hillier, N.; Arumugam, S.; Beeby, S. Screen Printed Flexible Water Activated Battery on Woven Cotton Textile as a Power Supply for E-Textile Applications. *IEEE Access* **2020**, *8*, 206958–206965. [CrossRef]

engineering proceedings

Proceeding Paper

Solution-Processed Organic Light-Emitting Electrochemical Cells (OLECs) with Blue Colour Emission via Silver-Nanowires (AgNWs) as Cathode [†]

Katie Court [1,*], Sasikumar Arumugam [1,2], Yi Li [1], Martin D. B. Charlton [3], John Tudor [1], David Harrowven [2] and Steve Beeby [1]

1 Smart Electronic Materials and Systems Research Group, School of Electronics and Computer Science (ECS), University of Southampton, Southampton SO17 1BJ, UK; S.Arumugam@soton.ac.uk (S.A.); Yi.Li@soton.ac.uk (Y.L.); mjt@ecs.soton.ac.uk (J.T.); spb@ecs.soton.ac.uk (S.B.)
2 Synthesis, Catalysis and Flow Group, School of Chemistry, University of Southampton, Southampton SO17 1BJ, UK; dch2@soton.ac.uk
3 Sustainable Electronic Technology Group, ECS, University of Southampton, Southampton SO17 1BJ, UK; mdbc@ecs.soton.ac.uk
* Correspondence: K.L.Court@soton.ac.uk
† Presented at the 3rd International Conference on the Challenges, Opportunities, Innovations and Applications in Electronic Textiles (E-Textiles 2021), Manchester, UK, 3–4 November 2021.

Abstract: Organic light-emitting polymers can be formulated into solutions that can be printed in ambient atmospheres and cured at low temperatures of <120 °C. The deposition techniques that can be employed include spin coating, spray coating and ink-jet printing. This provides the possibility of fabricating OLECs onto to a range of flexible substrates including textiles, hence enabling wearable electronics. In addition, the utilization of different polymers could produce light-emitting textiles in a range of colors. This work details the optimisation steps and challenges involved in the fabrication of OLECs on Indium Tin Oxide (ITO) glass prior to the transfer of the process onto a textile. A blue-emitting polymer Merck (NCMP) is used for the active layer and the device fabrication process is carried out at low temperatures in an ambient atmosphere. Working devices have been created on ITO glass to achieve the top blue emission with the next phase being the transfer onto textile. Blue LECs emission peak is captured at 520 nm with brightness of ~25 cd·m^{-2}.

Keywords: light-emitting electrochemical cells; spray coating; light-emitting textiles; solution processed

Citation: Court, K.; Arumugam, S.; Li, Y.; Charlton, M.D.B.; Tudor, J.; Harrowven, D.; Beeby, S. Solution-Processed Organic Light-Emitting Electrochemical Cells (OLECs) with Blue Colour Emission via Silver-Nanowires (AgNWs) as Cathode. *Eng. Proc.* **2022**, *15*, 21. https://doi.org/10.3390/engproc2022015021

Academic Editors: Kai Yang and Russel Torah

Published: 24 June 2022

Publisher's Note: MDPI stays neutral with regard to jurisdictional claims in published maps and institutional affiliations.

1. Introduction

The development of electronic textiles (E-textiles) has been a particular area of interest since 2010 among both academia and industrial researchers. The various e-textiles applications include: healthcare, the military, and wellbeing, and textile light emission within the visible wavelength is of significant interest due to the fact that it is typically used for communication and signal information [1].

A conventional approach to achieve light emission on textiles has been to incorporate light-emitting devices [2], for example, light-emitting diodes (LEDs), by attaching them onto standard textiles as a carrier. However, this approach has limited technological novelty and changes the original texture of the textile permanently. As a recent alternative approach, light-emitting yarn has been used as it can be woven into the textile during fabrication to achieve a built-in light-emitting textile [3]. This approach also has its limitations on physical deformation from the yarn geometry and limitation on the light emission from the wrap and weft orientation. Another recent approach is to integrate the individual or a number of LEDs into a single yarn, then to sew or weave it into textile [4]; however, this approach limits the light-emitting textile to a single point or a set of single-point light source. So far, the state-of-the-art fabrication method to achieve the light-emitting textile is to deposit, for example, via a solution process, each functional layer of the light-emitting device directly onto the textile. This approach is typically used in LEDs and light-emitting electrochemical cells (LECs) [5]. However, this process presents significant challenges and requires multi-disciplinary contributions from synthetic chemistry, material science, printing techniques and ink formulation.

In this research work, we report the early-stage experimental results of a blue colour-emitting material, based on an ITO-coated glass substrate with top emission. This achieved light emission and its fabrication process has been optimised for the migration from an ITO-coated glass substrate to a textile substrate. The key novelties arising from this work are to optimise the fabrication process to make it suitable for transfer to the standard textile, whilst having limitations in the maximum processing temperature of 120 °C. At the same time, top emission is required for the textile substrate, as bottom emission is not suitable for non-transparent nor translucent textile substrates. The first blue LECs based on conjugated polymers were reported by Pei et al. in 1996 [6]. LECs fabricated with the highly efficient poly[9,9-bis(3,6-dioxaheptyl)-fluorene-2,7-diyl] (BDOH-PF), exhibited bright sky-blue emissions, with a max brightness of 190 cd·m^{-2}, under 3.1 V. Edman and co-workers introduced a tri-layer device structure of LECs showing excellent performances, with blue emissions peaking at the 450 and 484 nm regions with high efficiency of 5.3 cd·A^{-1} at a low drive voltage of 5 V [7]. The authors reported blue emission on and via the ITO glass substrate previously [8], which led to this further research work, demonstrating the spray coating of AgNWs as a translucent top electrode for blue emission via the top electrode. To date, no such blue emission devices have been reported to be able to meet the above criteria for the fabrication process migration to textile substrates for future textile blue light emission.

2. Materials and Device Fabrication

Prior to the device fabrication, the blue colour-emitting active layer formulation was prepared by making three separate 10mg/ml in cyclohexanone (~99%, Gillingham, UK) master solutions of the blue light-emitting polymer (NCMP, Livilux, Merck KGaA, Darmstadt Germany); an ion-dissolving polymer, Trimethylolopropane ethoxylate (TMPE-OH, average Mn ~450 g/mol, Sigma Aldrich, Gillingham, UK); and a salt, potassium triflouromethanesulfonate (KTF, ~98%, Sigma Aldrich, Gillingham, UK). The solutions were prepared and heated with magnetic stirring until dissolved. They were then blended in a ratio of NCMP:TMPE:KTF of 1.0:1.0:0.2. The blend was left to stir overnight at room temperature to ensure it was homogenous before use.

The device optimisation and fabrication work were carried out in an ambient atmosphere on pre-coated patterned indium in oxide (ITO) glass substrates (sourced from Guluo Glass, LuoYang, China). The substrates were cleaned by subsequent immersion in acetone, iso-propyl alcohol and deionised water. After drying with compressed air, they were exposed to a 15 min ultraviolet (UV) Ozone treatment using a Novascan benchtop ozone cleaner (PSD-UV4 with OES-1000D, Boone, NC, USA) in order to achieve the correct surface energy for the following PEDOT:PSS (Clevious P VP AI 4083, Heraeus, Leverkusen, Germany) layer. The PEDOT:PSS solution was spin coated onto the cleaned ITO and annealed on a hotplate at 120 °C for 20 min. The active layer formulation, the preparation of which has already been described, was deposited on top of the PEDOT:PSS layer by spin coating; this was also annealed on a hotplate at 120 °C for 30 min. The top electrode was either sputter coated silver in a vacuum chamber at 2×10^{-2} mbar, or spray coated AgNWs (Dycotec, Calne, UK). In both cases, the electrodes were deposited through a shadow mask. Conductive paint was applied to the electrode contacts to achieve a reliable contact during measurement. The devices were then encapsulated with a UV-curable epoxy (Norland, NOA 61) and a cover slip in order to protect the cells from oxygen and moisture during the measurement process. Figure 1a shows a schematic of the OLEC architecture and Figure 1b shows an image of one of the devices fabricated with the silver nano-wires top electrode.

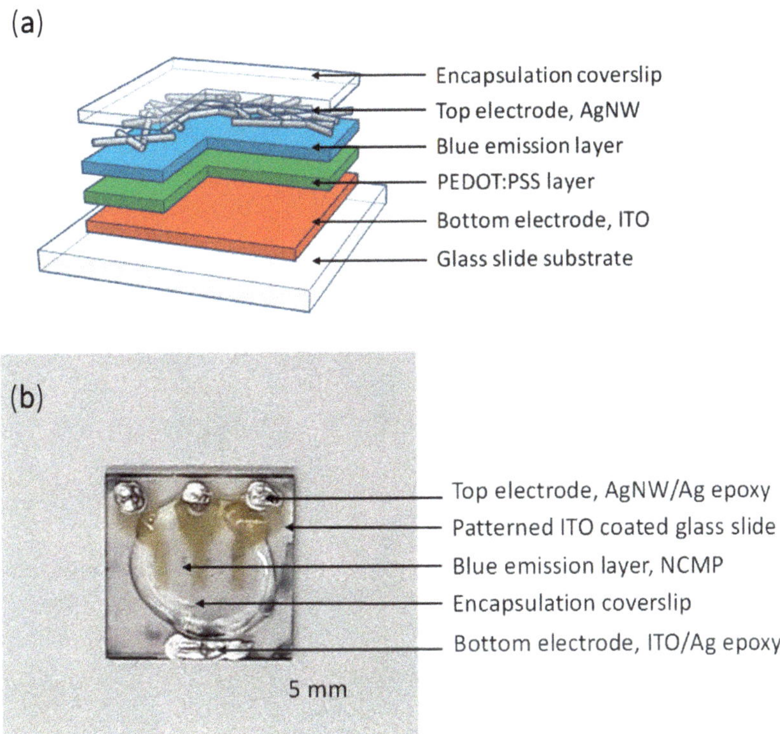

Figure 1. (**a**) Shows a pictorial schematic of the solution processed device architecture. (**b**) Image of the fabricated blue-emitting device.

3. Results and Discussion

Each step of the LEC fabrication process was studied in order to improve device performance and reproducibility. This has so far included optimising the coating and annealing conditions for both the PEDOT:PSS and active layers to yield layer thicknesses of 100 nm and 150 nm, respectively. The formulation preparation and ratio of components has also been studied as well as different post-fabrication annealing and encapsulation methods to improve the device lifetime during measurement. The reference device optimisation work used a sputter-coated silver electrode; then, the next stage of the process was studied with the solution processable top electrode formulations (AgNWs) in order to provide an entirely solution-processable process for when it is transferred onto a textile substrate. The optical properties of the blue emitter were also assessed, using an ultraviolet/visible (UV/Vis) spectrophotometer, and photoluminescence (PL) by laser excitation, as shown in Figure 2a, whereas Figure 2b shows the EL spectrum of the biased OLECs with the AgNWs top electrode, indicating the peak of 520 nm within the blue region. The peak of the emission corresponds with a wavelength of 520 nm, which corresponds to a brightness measurement of ~25 cd·m^{-2}. Figure 2c shows one of the fabricated devices (reference device) lit up with a bright blue light during measurement. The image of the AgNW lit-up devices is not available in this submission.

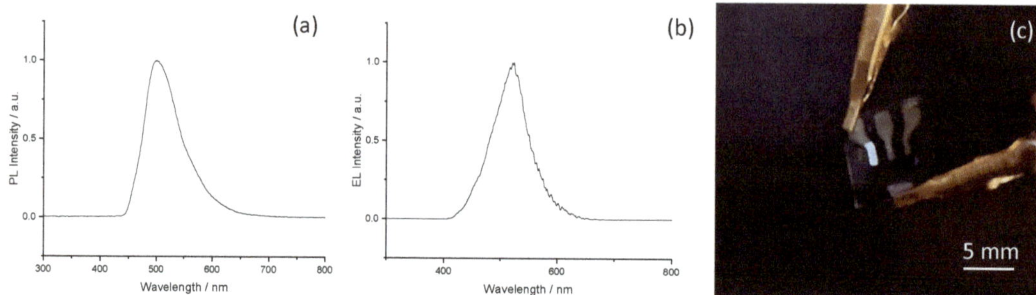

Figure 2. (**a**) Photoluminescent (PL) intensity spectrum obtained by laser excitation on the Merck NCMP in cyclohexanone solution, (**b**) electroluminescent (EL) intensity spectrum obtained by electrical bias at 6 V on the NCMP OLEC (spectrometer integration time of 500 µs), (**c**) the NCMP Merck blue-emitting device (reference device) lit up during measurement with a brilliant blue.

4. Conclusions

Blue-emitting OLEC devices have been fabricated using a process in which all steps were solution processed and in an ambient atmosphere. ITO pre-coated glass substrates were used for this initial optimisation work and the next steps will be to transfer the process onto a textile substrate. A commercially available blue-emitting polymer was utilized as the active layer, and by achieving the correct layer thicknesses for both the PEDOT:PSS and blue-emitter layers, bright light emission with good junction resistance (<1 kΩ) was achieved. A silver nano-wire spray coating formulation was used as a top electrode was used to achieve a solution-processed top electrode instead of the traditional sputter-coated silver. It can be seen from the image of the device that more work is needed to make more translucent cells for the top electrode for potential migration to textile substrates.

Author Contributions: K.C. and S.A. mainly did the lab work on fabrication; Y.L. did the measurement and characterization of the blue colour light emitting electrochemical cells. K.C., Y.L. and S.A. did the experimental plan, interpreted the collected data and prepared the manuscript. M.D.B.C. contributed his optical lab facilities for all the measurement. J.T. joined the technical discussion and proofread the proceeding. D.H. is the budget holder for the project. S.B. is the team leader and supervised the project. All authors have read and agreed to the published version of the manuscript.

Funding: This work was supported by the EPSRC Funding, Functional Electronic Textiles for Light Emitting and Colour Changing Applications (EP/S005307/1), Zepler institute Stimulus Fund 2017, Functional Electronic Textiles for Light Emitting Applications via Light Emitting Electrochemical Cells (517719107) and ERDF Interreg Program (EU), Smart Textiles for Regional Industry and Smart Specialisation Sectors (SmartT).

Data Availability Statement: Data available upon request, please contact the author.

Conflicts of Interest: The authors declare no conflict of interest.

References

1. Komolafe, A.; Zaghari, B.; Torah, R.; Weddell, A.S.; Khanbareh, H.; Tsikriteas, Z.M.; Vousden, M.; Wagih, M.; Jurado, U.T.; Shi, J.; et al. E-Textile Technology Review—From Materials to Application. *IEEE Access* **2021**, *9*, 97152–97179. [CrossRef]
2. Cinquino, M.; Prontera, C.T.; Pugliese, M.; Giannuzzi, R.; Taurino, D.; Gigli, G.; Maiorano, V. Light-Emitting Textiles: Device Architectures, Working Principles, and Applications. *Micromachines* **2021**, *12*, 652. [CrossRef] [PubMed]
3. Komolafe, A.; Torah, R.; Wei, Y.; Nunes-Matos, H.; Li, M.; Hardy, D.; Dias, T.; Tudor, M.; Beeby, S. Integrating Flexible Filament Circuits for E-Textile Applications. *Adv. Mater. Technol.* **2019**, *4*, 1900176. [CrossRef]
4. Hardy, D.A.; Moneta, A.; Sakalyte, V.; Connolly, L.; Shahidi, A.; Hughes-Riley, T. Engineering a Costume for Performance Using Illuminated LED-Yarns. *Fibers* **2018**, *6*, 35. [CrossRef]
5. Arumugam, S.; Li, Y.; Pearce, J.; Charlton, M.D.B.; Tudor, J.; Harrowven, D.; Beeby, S. Spray-Coated Organic Light-Emitting Electrochemical Cells Realized on a Standard Woven Polyester Cotton Textile. *IEEE Trans. Electron. Devices* **2021**, *68*, 1717–1722. [CrossRef]

6. Pei, Q.; Yang, Y.; Yu, G.; Zhang, C.; Heeger, A.J. Polymer Light-Emitting Electrochemical Cells: In Situ Formation of a Light-Emitting p–n Junction. *J. Am. Chem. Soc.* **1996**, *118*, 3922–3929. [CrossRef] [PubMed]
7. Tang, S.; Sandström, A.; Fang, J.; Edman, L. A Solution-Processed Trilayer Electrochemical Device: Localizing the Light Emission for Optimized Performance. *J. Am. Chem. Soc.* **2012**, *134*, 14050–14055. [CrossRef] [PubMed]
8. Arumugam, S.; Li, Y.; Court, K.; Piana, G.; Charlton, M.; Tudor, J.; Harrowven, D.; Beeby, S. Solution processed blue light emitting electrochemical cells fabricated and encapsulated fully in ambient environment. In Proceedings of the 2021 IEEE International Conference on Flexible and Printable Sensors and Systems (FLEPS), Virtual, 20–23 June 2021; pp. 1–4.

engineering proceedings

Proceeding Paper

Functioning E-Textile Sensors for Car Infotainment Applications †

Pouya M. Khorsandi *, Alaa Nousir and Sara Nabil

iStudio Lab, School of Computing, Queen's University, Kingston, ON K7L 2N8, Canada;
21amra@queensu.ca (A.N.); sara.nabil@queensu.ca (S.N.)
* Correspondence: 20pmk2@queensu.ca
† Presented at the 3rd International Conference on the Challenges, Opportunities, Innovations and Applications in Electronic Textiles (E-Textiles 2021), Manchester, UK, 3–4 November 2021.

Abstract: Car interiors are envisioned to be living spaces that support a variety of non-driving-related activities. Previous work focuses on enhancing driving-related functions, performance and safety. By developing textile-based sensors, we focus on enabling non-driving activities integrated in the car interior and supporting a richer user experience. In this paper, we introduce an array of new applications using e-textile sensors to the design space of car interiors. Our functional prototypes implement hand interactions (such as press and double tap gestures) on the leather or fabric of the steering wheel and back of the head rest. We then propose applications for these sensors to control media, car windows, and air-conditioning. Overall, the paper contributes a novel tactile input modality to support drivers and empower backseat passengers.

Keywords: human-computer-interaction; human-vehicle interaction; e-textiles; user experience; non-driving activities; car interiors

Citation: Khorsandi, P.M.; Nousir, A.; Nabil, S. Functioning E-Textile Sensors for Car Infotainment Applications. *Eng. Proc.* **2022**, *15*, 22. https://doi.org/10.3390/engproc2022015022

Academic Editors: Steve Beeby, Kai Yang and Russel Torah

Published: 18 July 2022

Publisher's Note: MDPI stays neutral with regard to jurisdictional claims in published maps and institutional affiliations.

1. Introduction

In the past, in-car interaction modalities have been restricted to traditional mechanical gear, knobs, buttons and handles. Lately, graphical user interfaces (GUIs) were introduced to commercial cars, including dialogue-box representations and speech-based input. Today, novel technologies create many opportunities for designing valuable and attractive in-car user interfaces. For instance, designing for a richer experience is starting to shape technologies such as automated driving, creating unprecedented opportunities for designing further comforting and entertaining in-car user interfaces. For instance, designing for a richer experience is starting to shape technologies that assist the user in driving, such as navigation systems or voice assistant systems, where the user interface is essential to the way people perceive the driving experience. Scholars consider that in the near future cars will transform into living spaces [1] rather than just means of transportation. In that sense, the interior designs of cars should consequently be revisited, with interactivity being embedded ubiquitously within the interior fabric itself. Therefore, our motivation and inspiration in this project stems from the 'interioraction' concept [2] that supports the blend of both interaction design and interior design into a seamless union. New means for user interface development and interaction design are required, as the number of factors influencing the design space for automotive user interfaces increases [3]. This paper discusses novel interactions through parts of the car consisting of textiles such as fabric or leather, which constitutes a great potential opportunity to utilize these parts of the car for human–vehicle interactions. In the rest of the paper, we first discuss some of the existing tactile user interfaces for in-car systems in the Background section. Second, we mention some of the recent advances in e-textiles that can be incorporated in cars for in-car interactions, and finally we explore some of the possible applications of e-textiles in cars regarding non-driving activities and how e-textiles can potentially improve user experience.

2. Background

To enhance user experience inside the vehicle, various in-car systems have been designed and developed to fulfill the requirements of passengers during the journey, such as, for example, navigation systems, media players and multi-functional displays. As we know, all of these in-car systems in a manually-driven context are distracting and focus-demanding; thus, it is in opposition to the primary task of driving, which requires extensive visual and cognitive attention. As a result, engaging a driver with various auditory and visual demanding systems could cause devastating crashes [4]. To redeem this issue of distracting systems, many researchers have proposed to increase the modality of input and output channels so that drivers do not have to divide their visual attention between the primary task and in-vehicle systems.

One of the less employed interaction methods in vehicles is a tactile sensory modality that does not interfere with driving [5]. Many papers have discussed the effects of this modality on driving performance and user experience [4–6]. The results illustrate that there is no significant difference between the tactile display and conventional systems (audio/visual-based) in terms of performance; however, the user experience has improved: participants considered getting feedback from the tactile display to be comfortable and pleasant [4]. The enhanced interaction style was the result of an improvement in the usability and ease of use goals, not in the experiential values [7] such as the playful, engaging, aesthetic and pleasant aspects of the user experience.

Derived from the literature, the usage of textiles inside cars for in-car interactions is confined to the integration of tactile feedback (e.g., push, shear force) into the steering wheel or seat for navigational cues or notifications [8,9]. These tactile-based interfaces were only confined to drivers for assisting driving-related tasks. By using e-textile fabrication methods, we can design interfaces not only for drivers but also for 'passengers', who have been predominantly neglected when designing in-car interactions, to enhance their user experience in terms of experiential values and offer an excellent opportunity to embed seamless, less focus-demanding, playful interactions by employing e-textiles techniques.

3. Sensing and Actuation

With the rapid advance of e-textiles in recent years, we can transform textile components into interactive fabric that can receive inputs and give outputs without being integrated with bulky electronics.

Advancements in different kinds of e-textile input sensors such as pressure-sensitive textile sensors, conductive threads/fabrics, and advanced fabrication methods make it feasible to fabricate textile capacitive/resistive touchpads [10] and deformable textile sensors [11] to detect surface and deformation gestures efficiently and reliably. As for textile interfaces for output modality, there has been extensive progress on light-emitting textiles, e.g., the integration of inorganic printed LEDs [12], Ultraviolet Organic Light-Emitting Electrochemical Cells (UV OLECs) [13], fabric audio speakers [14] and shape-changing fabrics [15], to embed visual feedback through the textile for in-car interactions. Using these technologies, many in-car interactions are practicable through e-textiles, and we explore some of them in the next section.

4. Proposed Applications

As we transition towards automated cars and they are becoming a space for many entertainment and non-driving activities, more user interfaces are being added to the vehicle interior. As a result, this number of interfaces may be disruptive and sometimes annoying. Because of their pervasive nature in our daily lives and their vast presence in the car interior, in-car textiles offer us this great opportunity to morph the car interior into an interactive interior using e-textile disciplines. By reducing interfaces in the center stack and embedding them in the interior by utilizing 'interioractive' design considerations [2], the car interior would have many resemblances to a living space and could be more dynamic and adaptable to user needs, which could stimulate more pleasurable in-car interactions that could benefit passengers/drivers using manually driven cars and automated cars in terms of user experience.

4.1. Steering Wheel E-Textile Interaction

In manually driven cars, drivers are supposed to concentrate on the roads; therefore, interactions with in-car functions should require as little visual attention as possible. Considering this fact, the steering wheel center would be a perfect space for eyes-free interactions with the fabric of the steering wheel for non-driving tasks. In Figure 1, the proposed intuitive gestures for interacting with the media player, window and air conditioning through the center of the steering wheel have been proposed. According to Pfleging et al. [16], who embedded touch screens in the center of the steering wheel to capture in-car functions, the most repeated gestures were sliding up/down or up/down. Their results suggest that these gestures are easy to execute and remember.

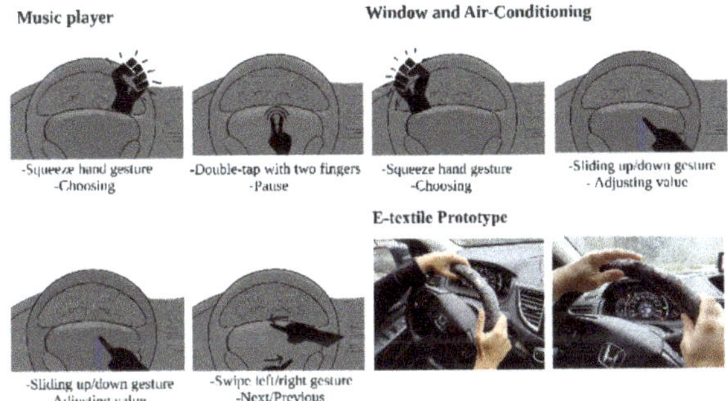

Figure 1. Illustrated gestures that can be executed on a steering wheel to control a media player, window and A/C, and the fabricated e-textile prototype.

Based on previous work, we propose utilizing sliding gestures on the fabric to adjust the value of the three in-car systems: media player, windows and air conditioning. To choose which system one intends to interact with, voice commands could be used for activation, while hand gestures are less cognitively demanding in terms of learnability, which is more satisfactory. The hand squeeze gesture is also an intuitive gesture that can be executed for activation. The steering wheel can be divided into four sections, and each section corresponds to a specific system. An e-textile prototype for the steering wheel has been fabricated and is depicted in Figure 1 as a proof of concept.

This prototype was developed using Knit Jersey conductive fabric from Adafruit and a sewing machine for stitching the conductive fabric to the base fabric. The conductive fabric is connected to the Lilypad microcontroller board through soldered silver-plated conductive thread from Karl–Grimm stitched on the conductive fabric as seams. HID (Human Interface Devices) control keys for controlling the multi-media player are sent to the media player using the Adafruit Bluefruit LE UART Friend module attached to the Lilypad board after detecting the executed gesture on conductive fabrics.

4.2. Headrest E-Textile Interaction

Since car interior textiles are within the range of passengers' hands, they can be a new modality for in-car interactions. The back of the headrest is one of the areas in cars that is mainly used for touch screens, and to the best of our knowledge, no prior work has explored this space as an opportunity for embedding e-textiles and transforming it into interactive fabric for in-car interactions, especially for passengers. E-textiles enable us to integrate digital capabilities behind and/or within fabrics and keep the car interior as seamless as possible, allowing interior designers of future automated vehicles to play with possibilities without designing encompassing screen-based interfaces. In Figure 2, the proposed gestures for passengers to interact with the three systems of the media player, window and air conditioning are illustrated. Unlike steering wheel gestures designed for drivers, headrest gestures designed for backseat passengers can utilize their visual channel. Visual feedback can also be possible with the usage of thermochromic threads or LED/UV OLECs. In Figure 2, the possible interactions are illustrated and the e-textile prototype has been embedded into the back of the headrest as a proof of concept. The e-textile interface can be enhanced to detect various gestures with embedded visual feedbacks. The fabrication and development process for this prototype is the same as for the previous prototype, with the exception of its visual design; here, it is designed for passengers' in-car interactions to control a media player.

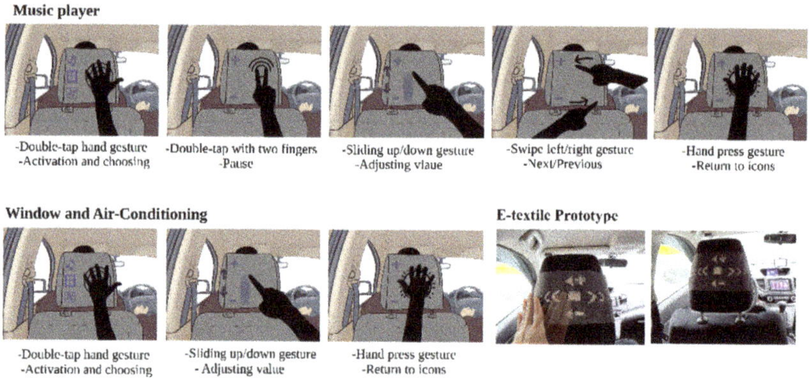

Figure 2. Illustrated gestures that can be executed on the back of the headrest to control a media player, window and A/C, and the fabricated e-textile prototype.

5. Conclusions

This paper discusses how e-textiles could be integrated into the human vehicle interaction domain, particularly as an input modality for both drivers and backseat passengers—the latter not having been explored before. We introduced a number of novel applications of e-textile interactions seamlessly embedded within car fabrics or leather surfaces. On both the steering wheel and the back of the headrest, we showed our proposed e-textile input interfaces as tactile screenless means of controlling multimedia, windows and A/C inside the car. By highlighting the gap between the e-textiles and Human-Vehicle Interaction (HVI) areas, we hope more research will be done in this area that evaluates how people interact with and perceive such tactile e-textile applications. As we transition towards fully automated vehicles (driving tasks will be removed from drivers, and all tasks will be non-driving tasks), more and more opportunities will arise for drivers' and passengers' in-car interactions through e-textiles. We hope to inspire future work around the situated use of such applications and their support of social engagement amongst family members or those with children and amongst strangers sharing or car-pooling, while also empowering backseat users using online taxi services.

Author Contributions: Conceptualization, P.M.K., A.N., S.N.; methodology, P.M.K.; writing—original draft preparation, P.M.K.; writing—review and editing, P.M.K., A.N., S.N.; visualization, P.M.K., A.N. All authors have read and agreed to the published version of the manuscript.

Funding: This research was funded by the Natural Science and Engineering Research Council of Canada (NSERC) through a Discovery grant (04135/2021), as well as through a Queen's University Research Initiation Grant (RIG).

Institutional Review Board Statement: Not applicable.

Informed Consent Statement: Not applicable.

Data Availability Statement: Not applicable.

Conflicts of Interest: The authors declare no conflict of interest.

References

1. Schartmüller, C.; Sarcar, S.; Riener, A.; Kun, A.L.; Shaer, O.; Boyle, L.N.; Iqbal, S. Automated cars as living rooms and offices: Challenges and opportunities. In Proceedings of the Extended Abstracts of the 2020 CHI Conference on Human Factors in Computing Systems, Honolulu, HI, USA, 25–30 April 2020; pp. 1–4. [CrossRef]
2. Nabil, S.; Kirk, D.S.; Plötz, T.; Trueman, J.; Chatting, D.; Dereshev, D.; Olivier, P. Interioractive: Smart materials in the hands of designers and architects for designing interactive interiors. In Proceedings of the DIS 2017—Proceedings of the 2017 ACM Conference on Designing Interactive Systems, Edinburgh, UK, 10–14 June 2017; pp. 379–390. [CrossRef]
3. Kern, D.; Schmidt, A. Design space for driver-based automotive user interfaces. In Proceedings of the 1st International Conference on Automotive User Interfaces and Interactive Vehicular Applications, AutomotiveUI, Essen Germany, 21–22 September 2009; pp. 3–10. [CrossRef]
4. Kern, D.; Marshall, P.; Hornecker, E.; Rogers, Y.; Schmidt, A. Enhancing navigation information with tactile output embedded into the steering wheel. In Proceedings of the International Conference on Pervasive Computing, Nara, Japan, 11–14 May 2019; pp. 42–58. [CrossRef]
5. Shakeri, G.; Brewster, S.A.; Williamson, J.; Ng, A. Evaluating haptic feedback on a steering wheel in a simulated driving scenario. In Proceedings of the 2016 CHI Conference Extended Abstracts on Human Factors in Computing Systems, San Jose, CA, USA, 7–12 May 2016; pp. 1744–1751. [CrossRef]
6. Kern, D.; Pfleging, B. Supporting interaction through haptic feedback in automotive user interfaces. *Interactions* **2013**, *20*, 16–21. [CrossRef]
7. Matsumura, K.; Kirk, D.S. On Active Passengering: Supporting In-Car Experiences. *Proc. ACM Interact. Mob. Wearable Ubiquitous Technol.* **2017**, *1*, 154. [CrossRef]
8. de Vries, S.C.; van Erp, J.B.; Kiefer, R.J. Direction coding using a tactile chair. *Appl. Ergon.* **2009**, *40*, 477–484. [CrossRef]
9. Asif, A.; Boll, S. Where to turn my car? Comparison of a Tactile Display and a Conventional Car Navigation System under High Load Condition Amna. In Proceedings of the 2nd International Conference on Automotive User Interfaces and Interactive Vehicular Applications, Pittsburgh, PA, USA, 11–12 November 2017; pp. 64–71. [CrossRef]
10. Ono, K.; Iwamura, S.; Ogie, A.; Baba, T.; Haimes, P. Textile++: Low cost textile interface using the principle of resistive touch sensing. In Proceedings of the ACM SIGGRAPH 2017 Studio, SIGGRAPH 2017, Los Angeles, CA, USA, 30 July–3 August 2017; Figure 2: 2–3. [CrossRef]
11. Parzer, P.; Sharma, A.; Vogl, A.; Steimle, J.; Olwal, A.; Haller, M. SmartSleeve: Real-time Sensing of Surface and Deformation Gestures on Flexible, Interactive Textiles, using a Hybrid Gesture Detection Pipeline. In Proceedings of the 30th Annual ACM Symposium on User Interface Software and Technology, Québec City, QC, Canada, 22–25 October 2017; pp. 565–577. [CrossRef]
12. Claypole, J.; Holder, A.; McCall, C.; Winters, A.; Ray, W.; Claypole, T. Inorganic Printed LEDs for Wearable Technology. *Proceedings* **2019**, *32*, 24. [CrossRef]
13. Arumugam, S.; Li, Y.; Pearce, J.; Charlton, M.D.B.; Tudor, J.; Harrowven, D.; Beeby, S. Visible and Ultraviolet Light Emitting Electrochemical Cells Realised on Woven Textiles. *Proceedings* **2021**, *68*, 9. [CrossRef]
14. Nabil, S.; Jones, L.; Girouard, A. Soft Speakers: Digital Embroidering of DIY Customizable Fabric Actuators. In Proceedings of the TEI 2021—15th International Conference on Tangible, Embedded, and Embodied Interaction, Salzburg, Austria, 14–17 February 2021. [CrossRef]
15. Nabil, S.; Kučera, J.; Karastathi, N.; Kirk, D.S.; Wright, P. Seamless seams: Crafting techniques for embedding fabrics with interactive actuation. In Proceedings of the DIS 2019—2019 ACM Designing Interactive Systems Conference, San Diego, CA, USA, 23–28 June 2019; pp. 987–999. [CrossRef]
16. Pfleging, B.; Schneegass, S.; Schmidt, A. Multimodal Interaction in the Car: Combining Speech and Gestures on the Steering Wheel. In Proceedings of the 4th International Conference on Automotive User Interfaces and Interactive Vehicular Applications, Portsmouth, NH, USA, 17–19 October 2012; pp. 155–162. [CrossRef]

MDPI

St. Alban-Anlage 66

4052 Basel

Switzerland

Tel. +41 61 683 77 34

Fax +41 61 302 89 18

www.mdpi.com

Engineering Proceedings Editorial Office

E-mail: engproc@mdpi.com

www.mdpi.com/journal/engproc